Management for Clinicians

A handbook for doctors and nurses

Tony White

Edward Arnold
A division of Hodder & Stoughton
LONDON MELBOURNE AUCKLAND

*To Simon, Daniel and Edward
for their humour and forebearance*

© 1993 Anthony White
First published in Great Britain 1993

British Library Cataloguing in Publication Data
White, Anthony
  Management for Clinicians
  I. Title
  362.1068

  ISBN 0–340–57321–x

All rights reserved. No part of this publication may be reproduced or transmitted in any form or by any means, electronically or mechanically, including photocopying, recording or any information storage or retrieval system, without either prior permission in writing from the publisher or a licence permitting restricted copying. In the United Kingdom such licences are issued by the Copyright Licensing Agency: 90 Tottenham Court Road, London W1P 9HE.

Whilst the advice and information in this book is believed to be true and accurate at the date of going to press, neither the author nor the publisher can accept any legal responsibility or liability for any errors or omissions that may be made.

Typeset in 10/12pt Palatino by Wearset, Boldon, Tyne & Wear
Printed and bound in Great Britain for Edward Arnold, a division of Hodder and Stoughton Limited, Mill Road, Dunton Green, Sevenoaks, Kent TN13 2YA by Biddles Ltd, Guildford and King's Lynn.

Foreword

In every western country the delivery of health care is both one of the largest employers of people and one of the largest consumers of gross national product. Even where it is a private rather than a public service, it is politically sensitive and attracts enormous local interest amongst the people who use, or will potentially use, its facilities. Health services are also organisationally complex. So it is hardly surprising that the management is highly problematic, and in many places underdeveloped.

This book addresses these problems directly. It rightly identifies that doctors, who make most of the decisions about consumption of resources, are central to health services being run efficiently and effectively. His book provides analysis and practical advice on how doctors and other managers can improve care for the population they serve.

David Costain
Director, Clinical Care Programme
King's Fund Centre

Preface

Doctors who find the introduction of the management philosophy into the National Health Service disturbing, unsettling or even threatening may be unhappy with phrases such as 'management' 'managing change' 'quality management' and 'corporate management'. Let me reassure you that this book is written by a clinician and is, I hope, based on a sensible and common-sense approach to the subject. It has been developed after wide consultation with clinicians, managers and academics in both the UK and the US.

The scientific basis of medicine leads to a reliance on scientific methods in organization. Doctors dislike social science, tending to place great validity on the outcomes of proper trials or demonstration projects rather than on post-positivistic methods. There are too many variables and no two situations are ever identical. This book is the result of a study largely based on cooperative inquiry, a new paradigm for research which breaks down the traditional distinction between the role of the researcher and the subject. This is succinctly described by Heron:[1]

> In the old paradigm only the researchers do the thinking that generates, designs, manages and draws conclusions from the research and only the subjects — often knowing nothing of what the researchers are up to in their thinking — are involved in the action and experience which the research is about. In the new paradigm this separation of roles is dissolved. Those doing the research as co-researchers are also involved as co-subjects. The same persons devise, manage and draw conclusions from the research; and also undergo the experiences and perform the actions that are being researched.

One of the striking, but not surprising, features and discoveries is that the same issue is seen so differently by differing groups. Doctors and managers frequently see problems in their own particular world as separate and therefore different. Indeed, it has been suggested that problems are not the problem of the organisation but rather the problem of the people as individuals. As Sims[2] states:

> ... problems are not 'things' with some external provable existence but constructions or definitions made by us in order to make sense of our world.

This leads on to the validity of the research and the methods used. For instance, the first response given to questions may not be the true thoughts of the subject but only the initial response given, without perhaps the opportunity for considered thought. The response may be influenced by the person asking the question. An answer that is considered to be helpful rather than accurate may be given. Indeed the research can be a tangled web of various influences and prejudices thay have to be taken into account.

I have loosely based my study on 'cooperative' inquiry but not the new paradigm for research explored by Reason and Rowan,[3] which breaks down the traditional distinction between the role of the researcher and the subject. It has been firmly bedded in the middle ground of ethnomethodology or participant comprehension. In collecting data from Clinical Directors and Managers from other units, and particularly from the United States, it would not be practical to use these new methods without thought. I therefore devised a method based on the concepts of the ethnomethodological approach but adapted it to fit the practicalities of the data collection.

One of the suggested explanations as to why clinicians are reluctant to become managers is the hypothesis that 'only poor doctors become managers'. A culture exists within the medical profession which, on the whole, does not recognise or acknowledge the value of management skills. Managers are accorded low status and this deters clinicians from assuming management roles. Career and professional disincentives, pressure of research and lack of reward for management activity also act as deterrents. They feel that a prime interest in patient care does not allow them time. In fact the medical professional has few relevant management skills and often feels disadvantaged when faced with experienced professional managers. These are all issues addressed in this book.

Consultants who hope that their involvement in management will diminish the importance of full time managers do not appreciate the complexity of running large organizations. Similarly, managers who assume they can achieve their objectives with disempowered consultants have lost sight of the main purpose of their work. What is needed, more than ever, is a comprehending partnership of the two that empowers both, for the benefit of the patient.

For a long time there has been increasing discontent and frustration by doctors trying to run clinical services in acute hospitals. They work long hours and feel there is little support from management. They are not sure what management could or would do for them, indeed management's purpose seems to them somewhat obscure. So much so that further involvement of doctors in management seems only likely to increase the consultants burden and deprive them of clinical freedom.

The dilemma, however, is that existing arrangements were unsatisfactory and failing. Medical committees and other so-called executive groups were in fact often working at best as advisory groups or merely as talking forums and providing safety valves for discontent. Neither could the problems of each consultant or faculty be represented to management through the chair or the minutes of these meetings.

It took the increasing crisis of funding and cost containment in the late 1980s to bring matters to a head. There was a realisation that a large proportion of the expenditure of an acute hospital was on the direct orders of doctors. By involving them in the decision-making process maybe costs could be contained.

Many have argued over the years, certainly since the Griffiths report, that the health service needs more clinicians to take a greater part in management. *Working for Patients* included proposals for ensuring consultants were involved in the management of hospitals and given responsibility for the use of resources. Consultants now have the opportunity to become involved in hospital management and should be encouraged to grasp this opportunity in order to re-establish their role in the management of hospitals. The changes taking place within the hospitals are really only just beginning.

Individual consultants see themselves as dedicated professionals providing high quality medical care yet sadly burdened with endless administrative chores, which often interfere with proper patient care. Often, there are no resources for the administrative work. The UK has a quality of care at least as good as any in other developed countries and achieved at lower cost than most. However the NHS overall projects a picture of a second rate underfunded service with inefficient administration, impersonal staff and dissatisfied patients suffering long delays for treatment. In the US most staff think highly of their hospital and are proud of what they achieve, the patients becoming caught up in the enthusiasm. The UK needs a more positive attitude to improving the public image and perhaps the more market-orientated environment, which was set out in the White Paper (1989), is designed to reward the efficient user of resources.

People are motivated by money, status, opportunity and, not least, by feeling valued and respected. People also need to feel part of a team, and a successful team. Leadership is usually the combination of attributes of a whole team. We need to develop a fully participative team culture. A little praise can make up for a great deal when working conditions are poor, with long hours and inadequate pay. These are difficult to endure if effort is unappreciated. People need to be recognised for their achievements and rewarding those achievements is an essential part of management skill. This is unfortunately sadly lacking in the NHS.

The NHS needs good leaders, both doctors and lay managers; innovators who can cut through red tape and bureaucracy and are positive towards problem solving. We need to encourage initiative and risk taking. We need to educate, train and appoint better managers, who respect their subordinates, who care about protecting the self-esteem of colleagues and subordinates during change and who consider it important to bolster confidence in staff and restore pride in achievement. People need to be motivated but you do not motivate people by talking about costs, neither are people motivated by talking to staff about quality issues. Mission statements do not motivate people. Managers (both doctors and lay professional) need to build effective teams. They need to ask the right questions of change, plan the implications for staff, train staff and communicate effectively. That does not mean issuing memoranda and instructions, nor copying every document to all staff or departments, but involving people and talking to them. The health service urgently requires more effective management processes.

We need enlightened ways of rewarding people for their achievements. Incentives, opportunities for further training and education, and more flexible promotion. We need to introduce flexible working arrangements for those who do not wish to work full time in the service, allowing their skills and self-esteem to be available and preserved while maintaining their status within the organisation. We need to avoid the 'burnout syndrome'. We also need better, brighter more congenial work places, comfortable canteen facilities, decent food, creches and places to park. We also need improved public relations facilities. We need leaders with a new vision to look beyond the immediate problems, to the longer term. To draw on medical audit data as it emerges and to use it to achieve greater success rather than reduce resources. We need creative managers to review current work patterns and promote services. We need managers both clinical and professional who share a common vision and have the courage to act on it.

References

1. Heron J. Validity in co-operative inquiry. In: Reason P., ed. *Human inquiry in action*. London: Sage Publications, 1989.
2. Sims DBP. From harmony to counterpoint. In: Mangham, I.L., ed. *Organization analysis and development*. Chichester: Wiley 1987.
3. Reason P, Rowan J. *Human Inquiry*. Chichester: Wiley, 1981.
4. Department of Health and Social Security. Griffiths Report. London: HMSO, 1983.

Acknowledgements

I am indebted to all those clinical directors and managers who participated in the research and gave so willingly of their time. In particular Professor Ron Bailey, University of Texas Medical Centre, Ken Bloem CEO of Stanford University Hospital, Stephen Royal CEO John Seely Hospital, University of Texas, Professor Cyril Chantler, Guys Hospital. To Dr John Mitchel and Dr John McClenahan of the King's Fund College, Dr David Costain, The Director, The King's Fund Centre and Dr John Yates of Inter Authority Comparison Centre, Health Services Management Centre. Wessex Regional Health Authority and the Royal United Hospital, Bath. To my colleagues and particularly the staff of the Postgraduate Medical library for their unstinting work on my behalf.

To those who provided wisdom and guidance in particular Dr Janet Gale of Centre for Postgraduate Medical Education, Dr Rodney Gale of Rodney Gale Associates, Dr Gifford Batstone Wessex Regional Audit Co-ordinator, Dr Jenny Simpson of BAMM and Mr Tim Scott of Health Strategies.

I am particularly grateful to my tutor Dr David Sims, Department of Organizational Studies, School of Management, University of Bath for his confidence in me, his enthusiasm and encouragement.

I freely acknowledge that some ideas may have been seeds sown by any of these people during the hours of discussions. I apologise if any reference is not attributed or incorrectly acknowledged, any such errors are mine alone.

I am greatly indebted to Anne Cooley for her valued chapter on nursing aspects. I could not have wished for a more enthusiastic and efficient contributor, always willing to meet deadlines.

To my editor Diane Leadbetter-Conway at Edward Arnold for her patience in the production of this book.

Finally to my wife Anne and my sons without whose inspiration, support, tolerance and encouragement the whole project would not have been possible.

Contents

Preface	iv
1 Doctors' roles in management	1
2 Models of clinical management	19
3 Managing as clinical director	40
4 Nursing management issues	59
5 Managing audit	78
6 Managing change	96
7 Managing meetings	114
8 Managing the Chair	132
Index	145

1

Doctors' roles in management
Tony White

Introduction	1
Visions of the future	4
Developing management skills	7
Involvement in the management of resources	13
Appointment of doctors to managerial positions	15
Reasons for involving doctors in management	16
Clinical freedom	17
References	17

Nos numerus sumus et fruges consumere nati. (Horace)

Introduction

It has always seemed difficult to involve clinicians in the problems of management, yet it is important that they should be involved. There are, however, very few incentives to encourage doctors to become more involved in management outside their own group or department. There is also a cost in terms of increased workload, financial loss and tensions with colleagues. The NHS Management Inquiry,[1] known as the Griffiths Report, was one of a series of central initiatives that aimed at involving doctors more effectively in management, but in reality offered little advice as to how that could be achieved. One of the main issues is the difficult relationship between doctors and managers. What can be done to improve matters? One way would be to explore and understand the tensions existing between doctors and general managers, some of which are historical, others related to differences in training, the different contexts of their work and the cultural differences. Handy[2] used Greek mythology to symbolise the different styles of management and culture found in organisations. Apollo, the god of order and bureaucracy, the patron god of the role culture based not on personalities but on definition of the jobs to be done, and Dionysus, the preferred god of artists and professionals within the existential culture, people who owe little or no allegiance to a boss:

Dionysians recognise no 'boss', although they may accept co-ordination for their own long-term convenience. Management in their organisation is a chore, something that has to happen like housekeeping. And like a housekeeper, a manager has small renown: an administrator amongst the prima donnas is bottom of the status lists.

Managers suffer pressure from the centre to ensure that regional and national priorities for care are duly met and that resources are used efficiently, while at the same time meeting local needs and the aspirations of consultants to provide increasingly expensive hi-tech medical care. In addition they suffer the frustrations of having to meet these responsibilities at a time of financial constraint, changing population needs, etc. and also recognise that they have no direct control over the major user of resources, i.e. the doctors. This division of responsibility has been inbuilt in the health service. Williams[3] states:

> An unfortunate division of responsibility seems to have grown up whereby costs are the business of administrators and treasurers, while benefits are the business of doctors and nurses.

Consultants have always had a major effect on the resources used. They respond to changes in demand for their services and to advances in medical technology. They make judgements about the skills and interests needed to replace retiring colleagues and are responsible for other major matters of policy. Developments that seem natural and proper to clinicians might take resources away from other parts of the hospital and may well create problems for future planning in other departments. Historically, many consultants have been concerned more with 'empire building' and preservation of their individual practices, often to the extent of inappropriate appointments of colleagues, than with wider issues of improvement in services to the public and appropriate use of resources.

Managers know that doctors affect managerial choices when they expand their operations and treatments without taking account of the resource implications. They also feel that a shift in clinical direction may owe more to personal interest than to the needs of the hospital. They find it frustrating that spontaneous actions by consultants can upset financial programmes for a district. Managers therefore face the difficult task of getting doctors to accept that their clinical freedom must be counterbalanced by an awareness of, and responsibility for, the effective management of resources. The task is made more difficult by the managers' lack of clinical knowledge, which may make them hesitate when discussing professional and technical standards and can make them feel vulnerable when issues are raised with doctors on quality improvement. Managers can find medical discussions direct and harsh to the point of sarcasm. On the other hand, doctors may find managers' conversations equivocal, facile and vague, and their thinking woolly.

Managers feel that they try to be careful in taking clinical issues into account in both the short and long term, through formal and informal consultation. They feel that arrangements are made for clinicians to influence and participate in management decision making at all levels. There is, however, a major dissonance between management and financial planning and clinical practice and developments. Both are legitimate but at times seem incompatible.

Doctors often fear management in its broadest term, and general mangers and chief executives are usually less aware than the doctors of just how suspicious and fearful the consultant body has become of general management. There are four main reasons for this although by no means inclusive of all reasons:

Erosion of power

Doctors have long feared that management, general managers and chief executives are trying to encroach on their long-held, professional independence — their freedom to determine their work loads and pattern of working and even their clinical freedom. They also fear that managers will remove them further from the decision-making machinery, thus leaving the mangers free to ignore medical advice.

Erosion of values

Doctors have always felt that health care is distinctive and special, characteristics that make it unresponsive and not appropriate for a managerial or commercial approach. Ethical values, it is felt, should not be tainted by a business mentality. Consultants often see discussions of resources as somehow improper, believing that such discussion conflicts with their responsibility to the individual patient. There is a fear that more management power may result in a financially led assessment of health care, which would force doctors to make decisions on economic rather than clinical grounds. Many words and phrases seem to have different connotations for the two groups. Phrases such as 'quality assurance', 'performance review', 'objective setting', 'monitoring', and 'efficiency' may be interpreted as tools for checking up on doctors or for cutting services. Doctors may see the use of terms like 'consumer' in place of 'patient', or 'marketing', 'public relations', or 'annual report', as evidence that health service values are being replaced by commercial ones. It is a mistake to introduce too many commercial-looking projects without adequate explanation. There are often mutual misunderstandings of the significance of particular words, and hence of each other's values.

Erosion of autonomy

Doctors often see managers as bureaucratic henchmen, in post to ensure that dictats from the centre are implemented. Individual performance review and performance-related pay has exacerbated those fears that a manager worried about an adverse personal review is more likely to act as an agent of the centre.

Cultural differences

Many of the difficulties in the doctors' responses to management stem from the fact that their cultural norms contrast so strongly with the managerial culture. This has already been alluded to in the reference above to the work of Handy[2]. Doctors are trained to do all that needs to be done for their patients, regardless of the effort or cost. They learn to be self-assured in defending their opinions and practices. They are expected to strive for the best available evidence before making a decision. They are used to working to short-term operational goals. Doctors rarely receive any training in management or organisational skills. They tend, therefore, to have a poor grasp of, or indeed little respect for, managerial skills or structures. They believe managers are there to oil the wheels, ensuring adequate facilities and equipment when they are needed. Doctors may misunderstand their role for advice and negotiation and therefore be ineffective in the medical advisory machinery.

Managers, in contrast, stress the virtues of interpersonal skills and of enlisting the cooperation of others. They are expected to subsume individual interests to those of the organisation. They are trained to be aware of the wider implication of any activity within the organisation and are expected to make optimal use of limited resources. Managers are normally trained to work to long-term goals, although this may not always be very evident in the health service, where managers are usually only in post in any one hospital for a comparatively short period compared to a consultant.

Of course, such descriptions about sets of cultural attributes can easily slip into stereotypes, and few doctors or managers fit neatly into these ideal types. Many of the medical profession have a constructive and sympathetic view of management and many managers respect the position of the doctors.

Visions of the future

Managers' views of doctors

Health service managers would like doctors to be able to sit down and look at the impact of their activities on the total health care of the district, and to see these activities in the context of the broader system, and not just of the pathology in front of them. From this point, it would be possible to make sensible allocations of resources, for example, what kind of doctor is needed? This would prevent general or orthopaedic surgeons or some other powerful group, for instance, diverting resources that would be better deployed for an extra radiologist or ophthalmologist.

Managers feel that it is important to involve doctors more effectively in management, although few have expressed very clearly what this means or how it might be done. The common aspirations are, however, to remove the barriers and mutual suspicions between doctors and managers and to accept that, even though doctors (not unnaturally) offer a countervailing perspective to management, they want to take the organisation in the same direction and pull together with managers and the rest of the hospital, accepting that managers need to match services to resources for the whole organisation and understanding what that means. Managers accept that doctors need to take a key part in management but this should be by participation and compromise, not by veto. Doctors should help managers make choices by realising the scarcity of resources.

Unfortunately not all managers share these views, and one extreme management view is that managers must curtail the excessive power of the doctors, curbing them in to ensure that district plans are achieved. On the other hand, some managers see the medical profession as a powerful organisation that should not be emasculated but carefully cultivated so that the doctors' skills, energies and intelligence are harnessed to tackle issues, or even to manage themselves more effectively. Doctors should be involved in corporate decisions at district level, as opposed to being advised on decisions but not making them. These latter views become easier for managers within units and departments. But it is at higher levels where it is it more important to manage doctors by enlisting their cooperation in management or encouraging them to work as managers.

Unsurprisingly managers show elements of all views, although varying in relative proportions, not necessarily in relationship to their training or background but perhaps related to their management style, the problems they face and the characteristics of the doctors with whom they work.

How managers can help doctors

Some managers have graciously acknowledged, with perhaps some surprise, the aptitude of some consultants for management and have admitted that they had under-rated consultants' abilities in the past. Unfortunately doctors, especially consultants, have a management style that they often do not recognise as such and that might well be styled as arrogant, authoritarian, paternalistic or pompous. The consultant should be a team leader, but it is doubtful whether many consultants understand the true nature of managerial leadership. There are a number of ways in which managers can help doctors fulfil their role in management.

First, by understanding the doctor's point of view and making sure that efforts that any doctor makes to improve a service has some direct and immediate effect on that doctor's own department. For instance, when consultants make efforts to improve the management of their service, either in terms of efficiency savings or income generation, it is vital that their department receives much of the benefit and they do not see it spirited away to a central fund. Without some incentive, disillusionment can quickly arise.

A common and often successful argument that plays on both the doctor's fears and needs has been to stress that doctors must either help to manage or be managed. That unless they become involved in managing their services it will be done for them, perhaps by people who are less aware of the issues as seen by the doctors on the ground. This ploy has been rather over used and is increasingly countered by the argument that only doctors can make many of the decisions and therefore this argument is seen as an empty threat.

Management information is a potentially useful tool in building up the role of doctors in management. But management credibility is damaged if the information is inaccurate. Even if the information is accurate, managers lose credibility by misunderstanding or even by failing to show that they do understand its clinical or epidemiological implications. Sometimes managers may assume incorrectly that doctors are easily able to interpret the information for themselves particularly financial information. Gross ignorance here can damage doctors' credibility in the eyes of management.

Some managers have demonstrated basic misconceptions about medical matters and have thereby allowed doctors to dismiss them as naive about health matters. One manager could not understand why notes from other doctors or hospitals relating to a patient's previous treatment might be valuable, 'Surely our own hospital notes are adequate?' Other managers have lost credibility by appearing unduly patronising or promising more than they could reasonably deliver.

Encouraging doctors to become more effectively involved in management is but one example of the problem of managing change successfully and this is an issue addressed in Chapter 6. If doctors are to be successfully involved in management it is important to recognise:

- The differences between the values of doctors and managers, their ideology and agendas.
- The different boundaries of the doctors and managers demesne.
- The two groups have very different characteristics, which are often misjudged, one by the other.
- The two groups need to coexist effectively.

How doctors can help managers

Management decisions sometimes look arbitrary or ill-advised to clinicians because they rest on an incomplete appreciation of clinical actions and developments. As a clinician, have you done all you can to ensure that management is informed and understands any changes in your current activity likely to have resource implications, your longer term developments or technical improvements that you plan to introduce? Do you keep yourself informed of management's intentions by talking regularly to management about their plans in areas that might affect your speciality? Do you receive the policy papers, for example proposals to develop or curtail particular services, strategic and operational plans budgets and other financial data pertinent to your work? Do you read them to keep informed and do you find them helpful? If they are unhelpful what suggestions have you made to improve them.?

Do you keep in touch with your colleagues with faculty and directorate meetings? Are these meetings effective in establishing clear channels of communication between management and clinicians? If not, what are you as an individual or group doing to make them more effective?

To help management, a doctor could:

- Keep management informed of any change likely to have resource implications.
- Keep informed of management's intentions.
- Read policy papers.
- Try to change unhelpful information.
- Keep in touch with colleagues.
- Make sure communication is effective.

Developing management skills

Clinicians are inevitably involved in managing people, departments and resources, whether or not they explicitly recognise this as a managerial role. As a clinician, you should review the extent to which you play such roles, how effectively you play them and whether you could do more to develop the necessary skills either yourself or in your colleagues and juniors. In the last decade or so there have been increasing attempts towards involving doctors in the management of clinical services in the US and the UK,[4] which requires two things to occur:

Managers to relinquish central control

Unfortunately this is a generic problem of managers. The classic bad manager is somebody who cannot devolve, who cannot give up control.

This is a particular problem in the health service because many managers already have difficulty devolving to other managers, so devolving to professional groups who they feel have no experience in management is even more difficult for them.

Doctors accepting managerial responsibility

In addition to teaching, educating and training doctors to be managers it is also necessary to organise a culture of understanding health care in the wider context and encouraging doctors and managers to work as teams.

Neither of these necessary prerequisites of the change anticipated have so far happened as widely as might have been hoped or expected. Duncan Nichol[5] said:

> ... that in some areas managers still had to win the respect of consultants. If there is no respect for the leaders they will not be followed. We have to build on that relationship and in some areas we have a long way to go.

The relationship between doctors and managers has always been delicate, often strained, often heated. The Government made it clear in the recent reforms that the Department of Health's doors were open to the ideas of managers, but closed to doctors' objections. To the BMA and others the message was clear, in Government circles a manager is more important. The problem is how to bring doctors back into a position of leadership within the NHS. According to Jeremy Lee-Potter:[6]

> In a well managed company the best, the brightest, the leaders, are the managers. In the health service the best, the brightest, the leaders, are the doctors. Good management is essential, but the primary work of the health service is the work of doctors and nurses.

The trend towards involving doctors in management in acute hospitals in not confined to the UK. Its origins are usually traced back to the events at the Johns Hopkins Hospital in Baltimore.[7] But there is also a long history of consultants being involved in management within the NHS, going back to the old Medical Superintendent role. Medical Superintendents had responsibility for all aspects of hospital administration at the beginning of the health service. These roles became less common and gradually lay or professional administrators took over the administration of the hospital. In 1954 the Bradbeer Report[8] proposed the idea of the Medical Administrator. The participation of hospital doctors in management and their contribution to the costs of running the hospital service was a major theme of the first reorganisation. Doctors were responsible for the use of resources with varying, but often considerable, degrees of autonomy. After discussion between the Minister of Health and the profession in 1965, the Joint Working Party on the Organisation of Medical Work in Hospitals was set up to discuss the NHS, and particularly to review the hospital service. The working party produced three reports,[9-11] which were known as the Cogwheel reports because of the logo printed on their covers. The first report recommended the creation of divisions of broadly linked specialities, to include consultants and junior medical staff who would be able to constantly review their services and methods of provision. With hindsight this could be an early ancestor of medical audit.

It was suggested that Cogwheel divisions were set up on a faculty or speciality basis, such as surgery, medicine, obstetrics, pathology, etc. Much depended on the size of the hospital and the size of the departments, it was obviously necessary in smaller hospitals and specialities to form groupings under one division, such as Gynaecology, Obstetrics and Paediatrics, or Otolaryngology, Ophthalmology, Orthodontics and Oral Surgery. The Chairs of each division were to come together, as representatives, in each hospital as a Medical Executive Committee, which would coordinate the work and views of the division and provide a link with nursing and administration. It was suggested that the sort of problems they might consider would include bed management and the organisation of outpatient and inpatient resources.

Most hospitals gradually implemented the scheme, and by 1972 the second report was able to identify the essential elements of an effective Cogwheel system and to report that, in large acute hospitals particularly, the system had been helpful in improving communications, the reduction of inpatient waiting lists and the progressive control of medical expenditure.

The third report clarified the role of Cogwheel systems in the then newly reorganised NHS, because the new emphasis of the 1974 reorganisation was the part played by multidisciplinary teams in integrated management. Whereas Cogwheel had been set up as a doctor-dominated, hospital-based arrangement, the third report suggested that it should continue to deal with issues where the agreement and action of hospital doctors was the primary need, while problems requiring strong collaboration between all professional groups, both within the hospitals and in community services should be the province of the district management teams (DMTs) and their health care planning teams. It would still be appropriate for Cogwheel systems to concentrate on efficiency issues, and it would be helpful for hospital doctors to see their clinical freedom in the context of team work and the necessity of sharing resources. Again, we see the foreshadowing of Medical Audit.

Cogwheel divisions have required a considerable amount of administrative support. Support for the Cogwheel concept was fairly general, and an alternative was difficult to find given the clinical autonomy that consultants claimed and largely possessed. The Royal Commission noted an impatience amongst medical staff with the seemingly inevitable delays intrinsic within consensus management and they supported an executive team at hospital level, which they thought would speed things along. The idea of unit management teams was endorsed in *Patients First*[12] and in a DHSS Circular[13] issued in 1980 on the new structure, but the involvement of doctors was somewhat ambiguously stated. It was not until January 1982 that clear directions were given on how clinical members were to be appointed to the DMT following the 1982 reorganisation.[14] The consultant was to be elected by the consultant body and the GP by all GPs in a district. This marked a change in some places, where previously the District Medical Committee (itself a representative body) had elected the DMT medical representatives. These representatives usually served for a limited period only.

Following the 1982 reorganisation, unit management teams were set up, usually a quartet of doctor, nurse, treasurer and administrator, but in some cases including a hospital consultant and a general practitioner. The role of these teams was not altogether easy to determine, nor was their corporate relationship to the DMT.

The 1983 Griffiths report proposals[1] recommended modification to this type of team decision making. It was hospital doctors' criticisms of consensus management that encouraged the Secretary of State to ask for the Griffiths report in the first place. The result — the appointment of general managers at district and unit levels — led the BMA to say that such a post should be held by a doctor. Many doctors were doubtful that filling the role would be practicable given their comparative, or total lack of management training and in consideration of their prime commitment to patient treatment, which allowed little time for a managerial role. The dilemma could not be easily resolved. Doctors need to be involved closely with the decisions about health care, but cannot spend too much time away from their patients.

To a certain extent the idea of a top doctor had been tried before with the Medical Superintendent in some hospitals and with the Medical Officer of Health in local health authorities prior to the 1974 reorganisation. Medical Superintendents' posts atrophied well before 1974, but the Medical Officer of Health was a highly influential officer in local authorities, whose work was widely appreciated. The holders of these posts did not find it altogether easy to adapt to the different management principles following the reorganisation and this has left community medicine in a somewhat ambiguous position. Is community medicine about the management of medical work or is it about the management of the community's health? Despite the claims of the Hunter report,[15] which tried to amalgamate both the managerial and clinical responsibilities, the Royal Commission felt that the community physicians' role in planning, health education, epidemiology and environmental control should be encouraged. This implied that the

Developing management skills

more administrative tasks should be undertaken by administrators.

It is now clear, however, that Griffiths was actually recommending delegation[16] rather than the general management structure that was imposed on the health service:

> One or two things I did not intend. Whilst my name at the time was primarily connected with general management I personally took this as shorthand for the introduction of an effective management process. I did not intend that the result should be yet another profession in the National Health Service to work in parallel with the other professions.

One of the aims of the White Paper (1989) *Working for Patients*[17] was:

> ... to ensure that hospital consultants — whose decisions effectively commit substantial sums of money — are involved in the management of hospitals; are given responsibility for the use of resources; and are encouraged to use those resources more effectively.

How this has been done in various hospitals has been an evolutionary process, with each hospital setting up its own organisational structure. Some, indeed, have made no real change from existing structures, except changes of name and/or personnel. However, the Government, certainly since the Griffith Report, has been attempting to involve doctors:[18]

> Since the NHS Management Inquiry of 1983 there has been a concerted effort, initiated by central government, to induce doctors to be more interested in (narrowly) costs and (more broadly) resource management.

Interestingly Pollitt *et al.*[18] state that a major factor is the difficulty of turning doctors into resource managers, and what has made the whole issue harder to resolve is that NHS managers have 'demonstrated scarcely more enthusiasm for such attempts than have doctors'. Before the National Health Service, when consultants were 'honorary' they knew the price of every ampoule of catgut and worked within a very small budget, so that providing health care within a constrained budget was something was very real to them. Before 1948 the role of doctors in cost-containment in hospitals was perfectly understood, because the hospital never had enough money, and had a Medical Superintendent who was responsible for running the place on a day-to-day basis and who knew quite clearly that the hospital must remain within the budget set by the Board of Governors. The health service has trained successive generations to the notion that the service is free, regardless of the ability to pay and regardless of the ability of the government to pay. And there are still staff who say 'don't talk about money, lives are at stake'. This latest reorganisation is forcing people back to the allocation of health care within fixed resources. The fundamental problem of the health service is the inability of any advanced health service to do everything possible to everybody's benefit. You can never fund yourself out of that problem and you can't manage yourself out of it either. That problem will always be there, but it does have to be managed.

Traditionally in the NHS, doctors have taken or been given, little or no responsibility for budgets. Neither have they been given or taken much interest in the resource implications and costs of their decisions. The doctors have been very bad managers. The profession neglected its responsibilities as far as management was concerned and neglected its responsibilities in cost-containment and in allocation of priorities for a hospital, as opposed to an individual's interest. Now, as a result, as we are facing life as it really is, the Government is cutting back, the country is cutting back and we are all looking at reality, and we may well be a little bit too late.

The early NHS hospitals had run very much like the model of a private hospital, or to some extent how US hospitals were run. In other words, the doctor was not really regarded as an employee of that institution (until 1991, consultants, except those in teaching hospitals, were generally an employee of the Regional Health Authority). They advised the managers of the hospital what needed to be achieved medically, and it was the task of the managers to do their level best to satisfy the needs of the doctors. For the doctor it was a comfortable system.

To some extent, from 1948 until the late 1970s the NHS ran in this sort of way, and each year more and more money was spent. Perhaps the increases were not as great as those happening elsewhere in Europe, or as much as doctors would have liked but, none the less, each year, more was spent. The years 1948 to 1978 were the years of plenty, relatively speaking.

Then came the years of famine, with cash limits in 1979 and cash planning in the early 1980s, so that the old ways of running hospitals simply no longer applied. There is nothing unique about the NHS. Every country in the world is facing exactly the same problem of not having the resources to achieve everything they would like to achieve, and therefore choices have to be made. What changes are made the moment you apply cash limits are the responsibility of doctors within the organisation. In a cash-limited health service, striving for equity, everybody has to be accountable for what they do. Doctors have to be involved in making those choices, and they cannot be involved unless they accept responsibility as well as authority. When you involve professionals in the management of any service, responsibility and authority have to be decentralised in equal measure and accountability has to be clear.

There are ways in which doctors can be involved in and feel to be part of management in acute hospitals:

- By involving them in the management of resources, which endeavours to make them more cost conscious.

- By appointing them to managerial positions.

Involvement in the management of resources

Experience from other countries shows that doctors have not regretted participating in clinical budgeting. One of the main advantages is the strengthening of the doctor's position when negotiating resources. It enables discussion on standards and quality of care to be part of that discussion and some consultants have found this to be an incentive. However, some managers paradoxically feel that giving doctors resource/management information would be manipulated by consultants and intensify their 'shroud waving' ability and not necessarily make them better team members. Generally consultants are unhappy with management interference in medical workloads. They frequently distrust the accuracy of information.

Clinical budgeting

There is a gap between clinicians, who make decisions about who and how to treat, and managers, who have the responsibility for controlling a budget and keeping within cash limits. This gap has had a major impact on resources use. Managers did not see it as their role to be involved in negotiating with clinicians how resources should be used. Nor, since the beginning of the NHS, have doctors wished to fill this role. Clinical budgeting suggested that plans should be agreed by clinicians in conjunction with service providers and finance officers. These plans should incorporate objectives for clinical activity, specifying the details of resources required into a financial statement.

Management budgeting

The Griffiths Report[17] emphasised the need to involve doctors more effectively in the management of resources. It pointed out what has been emphasised on many occasions, that it is doctors' decisions that largely determine how resources are used.

Griffiths argued that clinicians should accept the managerial responsibility that went with the clinical freedom. He recommended that health authorities should:

- Involve the clinicians more closely in the management process, consistent with clinical freedom.
- Encourage clinicians to participate fully in decisions about priorities in the use of resources.
- Provide clinicians with administrative support, together with strictly

relevant management information, and a fully developed management budget approach.

This meant that clinicians, already involved informally and implicitly, were to be in future involved formally and explicitly in financial management and decision making, and be responsible for those decisions. The proposals thus had profound effects on general managers, treasurers and other professional groups, especially nurses. The discipline of management budgeting also meant that doctors would be accountable for their actions to a manager who might not necessarily be a doctor, a move that would have little appeal to the profession. It was essentially, however, similar to clinical budgeting.

A review of management budgeting in 1986[19] concluded that it had failed to achieve its objectives, due in part to a failure to win the support and commitment from key personnel, the absence of clear management structures and the rapid speed with which it had been introduced.

Resource management

A new initiative was needed and, in 1986, it came in the form of Resource Management[19]. The new approach was to aim for greater medical and nursing involvement, with a focus on measurable improvements in health care through better use of resources. There was a recognition that nurses and clinicians needed to be more involved than hitherto. The architect of the scheme, Ian Mills said:[20]

> The resource management programme is principally about changing attitudes and encouraging closer team work in managing resources among patient care professionals and between such professionals and other managers.

Resource management was intended to provide accurate and useful information to clinicians about their practice and costs compared with colleagues in the same hospital, district or region. In a sense it was the forerunner of medical audit in that it sought to encourage doctors to review performance and improve standards of health care. While some consultants found this information interesting and even useful, some became anxious about comparisons.

There is no doubt that winning over clinicians to resource management takes time. Earlier initiatives failed because they failed to convince clinicians that they themselves had anything to offer. Indeed the DHSS used the words 'seriously antagonised' in 1986[19] and went on to say that 'there may be a case for suspending management budgeting development for the time being'. Devlin[21] said that:

... unrewarding for the clinician, a fact that management consultants agreed in private conversations. In a cutback situation the health authority is having to grab every penny it can and will only squeeze further, consultants who improve their output. I think management budgets by incentive is fraudulent unless the clinician is prepared to go home and rest when he has reached his target output — doing more, more efficiently, negates the savings the health authority is really out to achieve. Savings, not efficiency, is the real bottom line.

Resource management therefore emphasised the human relations, whereas management budgeting was finance-led; resource management was more concerned with making doctors more management-conscious and accountable for the resources used.

Appointment of doctors to managerial positions

The Griffiths Report[17] emphasised the need for doctors to assume managerial responsibility. And this may take place in two main ways:

- By appointing doctors to general managerial posts.
- By involving doctors in management within hospitals.

Appointment of doctors to general managerial posts

Few doctors are willing to take general managerial posts. A small number were appointed; in 1987, less than 8 per cent at regional and district levels and less than 19 per cent at unit level. They were often on a part-time basis, and the numbers have since fallen. A number of reasons have been given for this, including:

- Limited interest in management roles.
- Lack of training and ability in management skills.
- Lack of suitable financial rewards for doing so, as it would have meant a cut in salary.
- A feeling from colleagues that, in doing so, one had crossed to the other side if you were full time, and being part time was difficult.
- There were difficulties in maintaining a career by abandoning all or even part of your clinical work.

Doctors in management within hospitals

In hospitals, Clinical Directors or Clinical Chairs (the terms vary between hospitals) are responsible and accountable for consultants and other medical staff within a directorate. In a classic case the Clinical Director is

16 *Doctors' roles in management*

supported by a Business Manager and Nurse Manager (this is set out more fully in Chapter 2). One of the problems in the Health Service is that we tend to use the same word for a number of things and many hospitals now have Clinical Directors, all of whom vary about what they actually do and regarding power, responsibility and authority.

Reasons for involving doctors in management

The major reasons given for the need to encourage more doctors to become involved in management seem to be:

- Need to reduce costs.
- To save money.
- To reduce overspends.
- Generally to assist with financial problems, particularly of resource not matching demand.

Disken *et al.*[22] quote the reasons for moving into some kind of clinical management as including:

- Decentralisation and delegation.
- To complement rapid developments in information systems.
- Paving the way for new information systems.
- To break down barriers between professional hierarchies and groups.
- To improve the quality of clinical services to patients.
- To reduce the cost of high cost services.
- Bringing the consultants 'on board' as a group.
- To allow more explicit evaluation of clinical work and outcomes.
- Severe financial problems.

They go on to state:

> ... that in most units there was a combination of such reasons and the roll out programme for resource management has speeded up the process of implementing clinical management structures. Reducing costs appears to have been the primary motivation in very few instances.

However, as doctors commit most of the resource, it is felt that this control could not be achieved without their cooperation and involvement. The problem has always been that in clinical work consultants were not subject to management control but worked in a professional hierarchy with considerable clinical freedom.

Clinical freedom

Clinical freedom, sometimes referred to as clinical autonomy, is the freedom conferred on clinical staff by the nature of illness and the relationship between doctor and patient, and has always been thought to be a prerequisite of providing health care. Many, including doctors, will argue with the various definitions and it should not be confused with the dilemma of personal advocacy versus corporate responsibility. It is at the core of the provision of health care. If you do not have a patient, i.e. someone in need of care and treatment, there is no need for a health service. If you do not have a person who can diagnose and treat that patient there is no health service. This has always, certainly in the past, given clinical staff key certain organisational rights, including the right of medical judgement, the right of independent practice, prime responsibility of individual patient care and an authority to determine what effort the NHS will put into each patient.

This clinical freedom or autonomy presumes unmanaged status for medical staff. This has in the past always been accepted. Rowbottom[23] suggests that some professions have reached the stage of development that prohibits management by non-members. If this is accepted then a single apex hierarchy will not be able to direct and control professional staff. New forms of managing professionals need to be sought. Private firms, which may have to organise professionals, do not have this problem of clinical freedom. It is this clinical freedom that politicians and managers seek to control, mainly in an effort to control costs.

Various governments of different shades are constantly trying to control the uncontrollable by altering the structure, but although the management structure can be altered there is no certainty of making the NHS more locally responsive or any better managed or any more efficient. It is important to identify those factors that make this so. The NHS is multiprofessional and multistructured, with various professional hierarchies. There are autonomous groups of staff, and huge spans of control. Some authors have claimed that any change in the health service is illusory, because the professional staff are not affected. It is the turbulence created by changing management arrangements that creates an illusion of change.

References

1. DHSS. *NHS management inquiry*. London: HMSO, 1983.
2. Handy C. *Gods of Management*, 3rd ed. London: Business Books Ltd, 1991.
3. Williams A. *Medical ethics: health service efficiency and clinical freedom*. Nuffield/York Portfolio No. 2. London: Nuffield Provincial Hospitals Trust, 1985.
4. Disken S, Dixon M, Halpern S, Shocket G. *Models of Clinical Management*. Institute of Health Services Management, London: 1990.

5. Nichol D. Report in *BMA News Review*, 1991; **17**(1): 10.
6. *BMA News Review*. 1991; **January**: 7(1): 12.
7. Heyssell RM, Gaintner JR, Kues IW, Jones AA, Lipstein SH. Decentralised management in a teaching hospital. *N Engl J Med* 1984: **310**(22): 1477–80.
8. Committee of the Central Health Services Council. *The Internal Administration of Hospitals* (Bradbeer Report). London: HMSO, 1954.
9. Ministry of Health. *First report of the joint working party on the organisation of medical work in hospitals*. London: HMSO, 1967.
10. Department of Health and Social Security. *Second report of the joint working party on the organisation of medical work in hospitals*. London: HMSO, 1972.
11. Department of Health and Social Security. *Third report of the joint working party on the organisation of medical work in hospitals*. London: HMSO, 1974.
12. DHSS. *Patients first*. London: HMSO, 1979.
13. DHSS. *Circular HC(80)8, Health service development*. London: HMSO, *Structure and management*. 1980.
14. DHSS. *Circular HC(82)1, Health service development*. London: HMSO, *Professional advisory machinery*. 1982.
15. DHSS. *Report of the Working Party on Medical Administrators* (Chairman Dr RB Hunter) London: HMSO, 1972.
16. Griffiths R. *Audit Commission annual lecture*. June 12 1991.
17. DHSS. *Working for Patients*. London: HMSO, 1989.
18. Pollitt C, Harrison S, Hunter D, Marnoch G. The reluctant managers: clinicians and budgets in the NHS. *Financial Accountability & Management* 1988; **4**(3): 213-33.
19. DHSS. *Health services management: resource management (management budgeting) in health authorities*. Circular HN Health Notice (86) 34. London: HMSO, 1986.
20. Resource Management Feedback. *The Resource Management Initiative* 1987; **1**: August.
21. Devlin B. Second opinion. *Health and Social Services Journal* 1985; **February 7**: 165.
22. Disken S, Dixon M, Halpern S, Shocket G. *Models of clinical management*. Institute of Health Services Management, 1990: 5.
23. Rowbottom R. Professionals in health and social service organizations. In: Elliott Jaques, ed. *Health services*. London: Heinemann, 1978.

2

Models of clinical management
Tony White

Introduction	19
Clinical directorates	20
Models of clinical directorates	22
Setting up directorates	25
Practical aspects of directorates	26
Classification of clinical directorates	27
Cultural change	29
The doctor's responsibilities	30
The opportunities	30
Hospitals' responsibilities	30
Clinical director appointments	32
Relationships with other managers	34
The medical director	37
References	39

Salus populi suprema est lex. (Cicero)

Introduction

It is important to distinguish between the clinical management structure within the unit or department and the management structure within the hospital, district or trust. There are some important differences of emphasis and the structure within one does not necessarily reflect the structure within the other. The Trust Boards are for strategic management, whereas Management Boards are for the operational management of the hospital or trust. Discussion in this chapter is limited to doctors' involvement in the management structure. I shall make no attempt to discuss in detail the management structure above Management and Executive Board level.

A number of hospitals in the UK have now been experimenting with variations of the clinical directorate management structure. This is the latest in a series of efforts to secure the active participation of clinical staff in the management of the hospital:

Models of clinical management

> Doctors and managers must build up a relationship of mutual dependence if the NHS reforms are to work.

Duncan Nichol, Chief Executive of the NHS, made this plea at a seminar on clinical directorates.[1] He went on to say that there was a need to sustain the unique doctor–patient relationship and then build a new and mutually supportive doctor–manager relationship.

It is not my intention to proscribe an ideal management structure, particularly as there are so many variations and no one example is better than another, but whatever arrangement is chosen has to be that which is best suited to the particular hospital, given all the local factors. There are, however, certain principles that determine success or failure. One of the main dilemmas is that what hospitals think, and profess they have chosen, does not always bear any resemblance to what an outside observer can discern. There are examples from various acute hospitals both here and in the US and some of the factors that work for or against effectiveness with regard to particular models have been considered.

In the smaller hospitals there seem to be typically four, five or six directorates; some of the largest hospitals having up to double this number and there are hospitals with over twenty and at least one hospital is planning to introduce fifty or sixty directorates. In large hospitals, where there are between one and two hundred consultants, it is usual, although by no means universal to split directorates into subdirectorates or associate directorates. In this situation it is often the Clinical Directors who have difficulty in devolving their managerial control. The demands on the time and the qualities needed of a Medical Director are also somewhat different in dealing with a structure containing five to seven directorates than one containing four times that number.

Clinical directorates

There are obviously no hard and fast rules for splitting the hospital into clinical directorates and specialities. The size of the hospital and number of consultants in total and within each specialty are factors to be taken into account. The aim would be to produce groupings of similar specialities with roughly some equality of size. A major centre with a large specialist unit would probably have separate directorates for those units, but directorates of more than a dozen consultants are probably too large, whereas those with less than a half a dozen, too small. Having said that, a useful rule of thumb is that medical and support service directorates can be larger than these figures as they usually contain less disparate groupings. The activity and control of the surgical specialities where services and activities are somewhat different has a far greater effect on hospital income. A hospital with five clinical directorates, might, if it contained the relevant specialities, have the following:

(1) Surgery, including General Surgery, Urology, Cardiothoracic Surgery, Neurosurgery, Orthopaedics and Trauma, including Accident and Emergency.
(2) Medicine, including General Medicine, Cardiology, Neurology, Dermatology, Nephrology, Gastroenterology, Geriatrics, etc.
(3) Specialist Surgery, including Plastic Surgery, Otolaryngology, Orthodontics, Oral Surgery and Ophthalmology.
(4) Obstetrics, Gynaecology and Paediatrics, etc.
(5) Support Services, including Anaesthetics, Radiology and Pathology.

A system split into seven clinical directorates would usually be divided as follows:

(1) Medicine, including General Medicine, Cardiology, Neurology, Dermatology, Nephrology, Gastroenterology, Geriatrics, Mental Health, Psychiatry, etc.
(2) Surgery, including General Surgery, Urology, Cardiothoracic Surgery, Neurosurgery, Orthopaedic Surgery including Accident and Emergency
(3) Specialist Surgery, including Plastic Surgery, Otolaryngology, Orthodontics, Oral Surgery and Ophthalmology.
(4) Obstetrics, Gynaecology, Paediatrics, etc.
(5) Anaesthetics, Intensive Care, Theatres, etc.
(6) Medical Imaging, Radiology, etc.
(7) Pathology, including Histology, Chemical Pathology, Microbiology, Haematology, etc.

Two further subdivisions would separate Orthopaedics (including Accident and Emergency) from Surgery and Mental Health and Psychiatry from Medicine, thus creating nine directorates. There are, as stated, no hard and fast rules; there are differences and anomalies. Directorates of only one consultant do exist alongside directorates in the same hospital with more than fourteen consultants. These anomalies appear to exist for reasons of personality, history and lack of management control. There are many hospitals who, for local reasons, have managed their directorate structures in differing ways. For instance, where hospitals are split between two or more sites, certain specialities, even though relatively small in their own specialist hospitals or site, may have their own directorate as a result.

It is, however, possible to plan an average picture of the splits as the number of directorates increases (Table 2.1). Some large hospitals, with perhaps 200 consultants, may have more than 30 directorates, although these are grouped into about six or eight clinical directorates, with the remainder as associate clinical directorates. The Associate Clinical Direc-

22 *Models of clinical management*

Table 2.1 Average divisions of faculties in clinical directorates

3 Dir's	5 Dir's	7 Dir's	8 Dir's	9 Dir's	10 Dir's
	Spec Surgery	Spec Surgery	Spec Surgery	Spec Surgery	Spec Surgery
Surgery	Surgery	Surgery	Surgery	Surgery	Surgery
			Orth/Tr	Orth/Tr	Orth/Tr
	O/G/Paed	O/G/Paed	O/G/Paed	O/G/Paed	Obs/Gynae
					Paeds
Medicine	Medicine	Medicine	Medicine	Medicine	Medicine
				Ment Hlth	Ment Hlth
Support Services	Support Services	Radiology	Radiology	Radiology	Radiology
		Pathology	Pathology	Pathology	Pathology
		Anaes/ICU	Anaes/ICU	Anaes/ICU	Anaes/ICU

tors are responsible for running these in exactly the same way, except that the Associate Clinical Directors do not sit on the Board, but meet regularly with their particular Clinical Director, who reflects their views to the Board. It is of course possible to have every faculty/speciality represented on the Management Board and this certainly produces a very desirable flat management structure. Unfortunately it demands more of the Medical Director in time, effort and quality of management skills than is usually available. Attempts to run such a structure with a Medical Director without that expertise or time, can be disastrous. What is also to be avoided is a system with large numbers of full Clinical Directorates (in excess of about fifteen) with token representation on the Board but no structure for direct communication with that Board.

Models of clinical directorates

The range of model found in the UK and indeed US hospitals is infinite. At the one extreme is the true manager, the Clinical Director, Chief, Chair, Director of faculty (the terms vary) who is clinician leader of a care group or speciality and who is half manager, half clinician. This person is allocated, or negotiates, a budget with senior management and is responsible for spending that budget. They have two contracts, one a contract as a clinical consultant, the other as a manager and they are therefore properly placed in general line management, bound by the ethos of being a manager. The Unit Manager may manage staff other than their own immediate junior staff, nurses for example, although professionally they are accountable to the Senior Nursing Officer. This post is, of necessity, a part-time activity, as they are generally practicing clinicians. For them, therefore, the quality of support staff is vital. These Directors are responsible and accountable for the planning of the service, allocation of resources and all operational work including the specification of internal contracts, taking responsibility for service planning and service development and working alongside a speciality manager who is responsible for operational management. The clinical director is at the heart of the process and is the key decision maker. This draws the clinician into corporate responsibility, accountabilty and influence. The difficulty lies in combining time-consuming management responsibility with clinical practice. There are problems of time and motivation, to say nothing of the cost to the service of loss of 'hands-on clinical work'. At present there are few clinicians with the necessary management expertise.

At the other extreme is what used to be Chair of Division — a representative who has been labelled a Clinical Director but for whom nothing else has changed. A speciality manager agrees activity targets between the Unit General Manager/Chief Executive Officer and the clinicians. The speciality is then provided with full management support by the speciality manager, who is the budget holder. Management responsibility lies with the full-time manager and clinicians concentrate on clinical work, where they remain on the periphery and are consequently less directly accountable. Managers in this situation are much more at the point of managing professionals directly.

Before considering the various models it might be helpful to consider the traditional model with the following organisational structure:

- Individual consultants prescribe treatment and care for their patients.
- A Chair is elected from the consultants within the faculty or division.
- These Chairs sit on a Medical Executive Committee (MEC).
- The Chair of that committee is elected by that committee.
- That elected Chair sits on the Management Board.
- Nurses and other professional groups are managed within their own separate hierarchies, and each has separate functional budgets.

Exceptionally the Chair of the MEC may be a senior consultant who is not a Chair of a division or faculty but who has been appointed by a caucus of senior consultants from the hospital.

The Directorate Model differs in a number of important ways:

- The Clinical Director is accountable to either the Unit General Manager or Chief Executive or the Medical Director.

24   Models of clinical management

- Medical Representatives of each and every faculty or division may no longer be represented on the Management Board.
- The nurses and other core staff in each clinical service are managed by the Clinical Director who holds the budget for the service.

In practice these pure forms of either extreme are rarely seen, particularly the true directorate form. Most Clinical Directors do not feel accountable to the Unit General Manager or Chief Executive, although the same Unit General Manager or Chief Executive may feel the Clinical Director is accountable to him/her. In reality, doctors tend to feel more comfortable when reporting to another doctor and for this reason it is desirable that a Clinical Director should report to the Medical Director rather than the General Manager or Chief Executive. Rarely is the Clinical Director responsible directly for the service budget. Prime responsibility does not, of course, necessarily mean exclusive responsibility, although this is a distinction that many managers find hard to understand. Therefore, although a 'lot of furniture has been moved around the room', little in reality has changed in many hospitals.

It might be helpful at this stage to consider some specific models that have been published and achieved notoriety, particularly as they are often quoted as a model by which a particular hospital works. Rarely is this found to be an accurate picture of the structure working within that hospital.

The Johns Hopkins model

Here each functional unit is headed by a Functional Unit Director (clinician). Reporting directly to each is a nursing director and an administrator; the three work together as a management team. They are accountable for all direct costs associated with the operation of the unit, including services from other departments, such as laboratory, medicine and radiology. Costs that pertain to the operation of the institution as a whole, e.g. central personnel administration, security, accounting, billing and insurance are allocated to the functional unit. Each unit may use services such as housekeeping, dietary and maintenance, from central hospital departments, but the unit may also switch to other providers. Each functional unit has, of course, to operate within the general policies of the hospital. It was felt that management strategies directed by physician managers were more likely to be successful as they could influence the behaviour of their colleagues.

The Guys Hospital model

Clinical directorates were established, each being headed by a clinician assisted by a Nurse Manager and a Business Manager.[2] The Business

Managers were mostly chosen from professional hospital administrators but could be a nurse or other professional such as a Scientific Officer. Some directorates share a Business Manager. Management accountability is seen as very distinct from professional accountability.

Clinical Directors are not elected but appointed, on advice from colleagues, with regard however to their management capabilities, as seen by the Chair of the board and the District General Manager.

Responsibility was then decentralised to the directorates so that over 60 per cent of total staff report within the directorates. These comprise doctors, nurses, clerical and scientific staff, etc. Centralised outpatient appointment and management arrangements, admissions and management of waiting lists were dismantled, the responsibility being assumed by the individual directorates. Rules for bed borrowing were also established and the authority of the ward sister over the ward has been reintroduced, including management of the ward budget.

The British Medical Association model

The BMA[3] states that:

> Although the ultimate managerial responsibility for a unit lies with the accountable officer, the executive responsibility for the management, finances and other resources of the unit should be carried by a single management body which contains a significant number of senior medical staff as full members.

BMA model hospitals have a single management body on which all clinical directors sit as full members. This body has full executive authority for the management of the units including finances and resources. Clinical Directors elected by consultants in that speciality are responsible for certain personnel and peer review matters, such as medical audit, but cannot override the clinical judgement of colleagues.

Setting up clinical directorates

Doctors provide long-term continuity of management to balance the constant change of managers, which can occur every three or four years. For this reason managers do not necessarily think long-term. Rules need to be set out explicitly. Management retains power by not setting out the rules of the game. Decentralisation of the organisation is vital. Unfortunately a minority of General Managers do not want doctors to become more involved in management. There needs to be a changing of mind sets and boundaries. The clinical directors' power base is the consultant body of the directorate. If you neglect this you cannot influence management.

There is no holy grail of structure to run a successful hospital, but clinical directors need to lead the directorate, take a role in hospital planning, be an ambassador at corporate level and involve the body of consultants in the management process.

Once the decision has been made in principle to implement an organisational management structure based on clinical directorates it is essential to dismantle all existing representative and advisory machinery. There will continue to be a need for a forum (such as a Medical Staff Committee) for all consultants to discuss matters of interest and importance to them, reflect views, etc. but this must not be part of the managerial structure and must have no explicit or implicit managerial role. It is not possible to run an old Cogwheel divisional system alongside or mixed with a clinical directorate system. As Marx said, 'Parallel structures are a way of subverting the organisation'. It is important that the clinical directors meet with the directorate and, although staff committees are important in an open organisation, they have no role in the management structure. Separate committees for the purchase of medical and scientific equipment and Medical Advisory Committees to Purchasers are also parallel structures with pseudo-management roles that subvert the new clinical directorate management structure of the hospital or trust. The divisional system was also doctor-dominated and had no involvement with the nurses and other professionals except as observers. The clinical directorate system works with Nurse Managers, Business Managers and Heads of Departments and discusses business matters relating to the directorate. Indeed, some would take the view that the change in 'business' may require a redefinition of clinical freedom.

Practical aspects of directorates

The extent to which any particular system will work in a particular hospital is a matter for discussion and full consideration by the staff within that hospital. It would not be possible to impose any particular structure on a hospital, but certain principles are necessary if clinicians are to be properly and genuinely involved in management. This must include the decentralisation of authority and responsibility and the development of team work between different professional groups, which is discussed in more detail later. It is important that the management skills of the other groups, the business managers and nurse managers are also developed.

It is vital that the Clinical Director, or Chair, does not end up as a glorified middle manager taking the responsibility for reductions in services together with all the budget restrictions. This may well curb clinicians if they have a management contract for the general management component of their work as they can be instructed to do certain things as manager that they would never contemplate as a clinician. This would be a middle management role that does not effect the allocation of resources. This is a common phenomenon and is a picture of a Clinical Director who is really little more than the old style Cogwheel chairman, representative of his colleagues and not part of the management chain but, who has alongside him, a Business Manager who is part of the management chain; between them they manage the unit. The Clinical Director here does not carry budget responsibility and is not part of line management but, as the doctor representative of colleagues, negotiates with the Manager. The only new feature in this organisation is that the management structure is allied to the clinical structure. Indeed, Kennedy[4] raises concern that there may be serious flaws in this clinical directorate model, ultimately making it unworkable for consultants and bad for nurses and others. He challenges the clinical directorate concept by asking a number of questions about the management responsibilities of the consultants, whether Clinical Directors can successfully direct consultant colleagues and whether there is true representation. It is significant that where Clinical Directors or Chairs have been appointed little has changed.

So what should Clinical Directors be? They are certainly not directors in the sense of ordering people to do this or that. Hence the reason for some hospitals' and consultants' dislike of the term. Some call the role 'Chair', others 'Consultant in Administrative Charge', others 'Head of the Department'. However, the actual title is less important than its functions. What is important is that a review and evaluation of a departmental organisation should focus on the roles of the Clinical Director, Business Manager and Nurse Manager. Such a review should consider:

- Does the department make the best use of the skills and time of the Clinical Directors?
- Are the members of the department aware of the implications for the hospital of the decisions it makes.
- Is there a good sense of collaboration, cooperation and working as a team?
- Are decisions being made at the level of impact?
- Is there proper responsibility, accountability and authority?
- Is initiative, innovation and risk-taking encouraged or suppressed?

Classification of clinical directorates

Through an analysis of the clinical director, business manager and nurse manger roles in hospitals it is possible to see that although the various typologies are infinite, they may be classified for organisational reasons into three main groups. In trying to establish such a classification it is necessary to consider the range of managerial responsibilities, roles and tasks and how these are or can be shared between the medical and

28 Models of clinical management

non-medical managers at departmental or faculty level. The crux of the matter is the relationship between the roles and the tasks and responsibilities of the three key players. The relationship between these key players and who they report to is an important but none the less separate issue. I have identified three models which bear a striking resemblance to the various ways of installing car engines: the transverse, the V and the straight in line models:

Sharing or transverse model

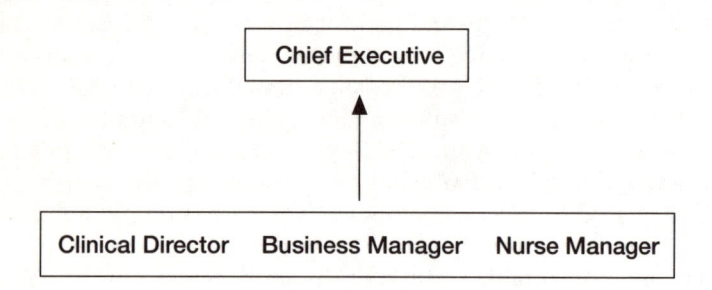

```
                    ┌──────────────────┐
                    │ Chief Executive  │
                    └────────▲─────────┘
                             │
┌──────────────────┬─────────┴────────┬──────────────────┐
│ Clinical Director│ Business Manager │  Nurse Manager   │
└──────────────────┴──────────────────┴──────────────────┘
```

The Clinical Director, Business Manager and Nurse Manager share the responsibility and accountability for both medical and non-medical decision making and report jointly to the Chief Executive. This enables the Clinical Director to be more comfortable reporting to a non medical authority.

Splitting or V model

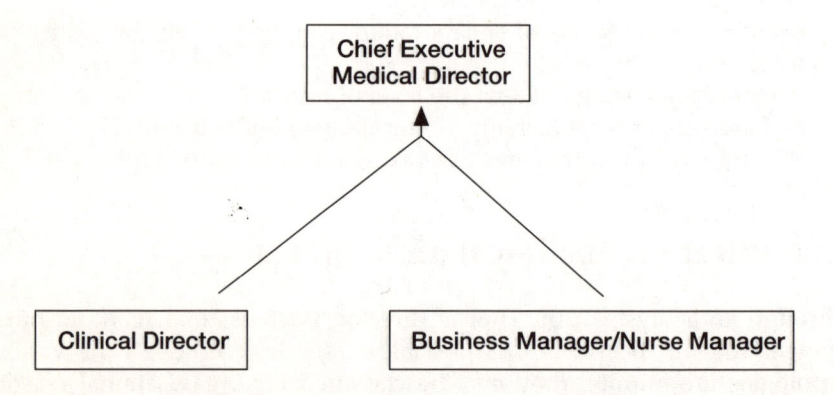

```
                ┌──────────────────┐
                │ Chief Executive  │
                │ Medical Director │
                └────────▲─────────┘
                        ╱ ╲
                       ╱   ╲
                      ╱     ╲
    ┌──────────────────┐   ┌──────────────────────────────┐
    │ Clinical Director│   │ Business Manager/Nurse Manager│
    └──────────────────┘   └──────────────────────────────┘
```

Here the organisation conveniently divides decisions into medical and non-medical, the Clinical Director taking responsibility and accountability for the medical decisions and reporting to the Medical Director as a representative of, rather than directly to, the Chief Executive or General Manager.

In-line or straight model

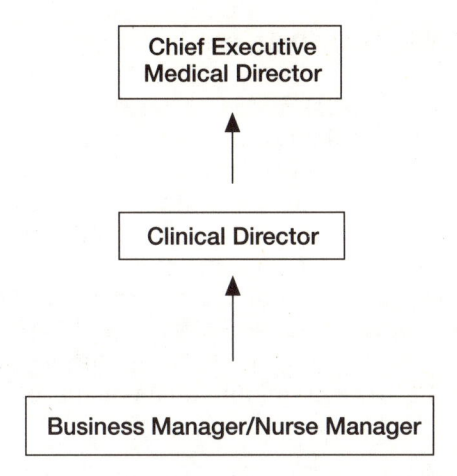

```
        ┌─────────────────┐
        │ Chief Executive │
        │ Medical Director│
        └─────────────────┘
                 ▲
                 │
        ┌─────────────────┐
        │ Clinical Director│
        └─────────────────┘
                 ▲
                 │
    ┌─────────────────────────┐
    │ Business Manager/Nurse Manager │
    └─────────────────────────┘
```

In this model the Clinical Director assumes total authority for the department; the Business Manager and Nurse Manger reporting directly to him. He in turn reports to the Medical Director as representative of the Chief Executive or Manager.

Cultural change

A change of culture is required to introduce a totally new and revolutionary management structure. Until now most management has been fire-fighting, with superficial changes and random initiatives. Resource management, clinical budgeting, awareness training and technical needs analysis are all efforts to create a sense of progress, but they are marginal. The culture of the hospital needs to be understood and managed dynamically and precisely so that the powerful and pervasive built-in inertias remain as they are and cannot gain strength and foil potential change. The time and energies of doctors, nurses and managers have in the past been diverted into battles rather than establishing commonalities, making the war unnecessary.

In the past, changes have been introduced that, on the surface, have produced change for managers to see, whereas below, in the body of the organisation, there is alienation and disaffection. Financial targets and activity achieved, but with low morale and minimal participation. Running the organisation below peak efficiency is wasteful and negligent.

The need for a strategy for cultural change is urgent. It needs to be understood, manoeuvred, adapted and prepared for the radical shift in management style. It is vital that UGM/CEOs address issues of cultural change and doctors attempt to understand the changes about to unfold.

The doctor's responsibilities

The establishment of Clinical Directors must have the support of the majority of medical staff as the organisational head, leader and figurehead of the directorate and responsible for the day-to-day operational and the strategic management of the directorate. Indeed, the Clinical Director may also be the line manager for the Nurse and Business Managers and therefore in contact with such management duties as objective setting and performance appraisal for these staff if this is required. Accountable for the management performance of those clinicians who have accepted budgetary responsibility within the directorate but excluding clinical responsibility for diagnosis and treatment of individual patients by other clinicians.

The opportunities

With clinical staff supported and empowered to make operational decisions, the hospital at a corporate level can then be freed to think clearly about strategic issues, direction and monitoring. New challenges and opportunities are opened up for, and welcomed by, many; the full potential of the hospital staff can be realised. Cost awareness can be seen and used as creating opportunity rather than inhibiting activity. Clinicians may then discover enthusiasm for management and an inherent ability. The freedom and power this creates should replace the authoritarianism of the old managed system. A re-evaluation of roles for both doctors and lay managers is necessary. Boundaries and traditional roles will change, people may certainly feel threatened but the change should produce benefits in greater democracy and awareness with considerable benefit to the patient as difficulties are confronted and addressed rather than ignored.

Hospitals' responsibilities

The BMA/Central Consultants and Specialists Committee recommend that clinical directorates, and indeed the constitution of a new management body, should be approved by the Medical Staff Committee. Consultant involvement in management has been a controversial issue. Some take the view that doctors should only practice medicine for which they have been highly trained, not 'dabble' in management, for which they have until recently had no training; 'I'm here to treat patients' is a familiar phrase. An alternative view is that management needs a strong clinical input and advice, without which inappropriate decisions and actions may be taken.

Equity

Emphasis has been placed on the need to avoid contracts or arrangements where priority for investigations and or treatment is on anything other than clinical need, irrespective of whether the patient is from a budget-holding general practice, health authority purchaser, extra-contractual purchaser or the private sector. The monitoring of this must be the responsibility of the clinicians.

Openness

Information should be freely available except in so far as it may relate to individual and identifiable patients or members of staff. The obligation in a publicly funded service is to public accountability, rather than commercial secrecy. There should be full explanations of decisions to relevant staff ahead of implementation (see Chapter 6) and, whenever possible, in time for amendments in the light of consultation. Weaknesses as well as strengths in the provision of clinical services should be identified and honestly acknowledged.

Bureaucracy

There should be continuing emphasis on reducing bureaucracy and simplifying the administrative process. One of the greatest problems in dismantling the existing management structure is that a bureaucracy is very difficult to dismantle because the bureaucrats within, resist and cling on to power.

Training

Management training should be provided as necessary, at present in many hospitals it falls between study leave and management expenses and in the end nobody pays. Continuing education is an area where for many years the health service has turned a blind eye and neglected the responsibility and welfare of the staff.

Clinical director appointments

Procedures

The appointment may be by internal choice of colleagues or the external appointment by UGM/CEO. To some extent the choice may be limited by those willing to undertake the role. There are important differences between the two. The representative can have more problems with, and complaints from, colleagues in difficult decisions because he is a representative, rather than an appointee, and the basis of the roles are different. Because of the need for accountabilty it is probably essential that the clinical director is appointed, but with the support and the confidence of the consultants within the directorate. It is important for a directorate not to put a 'wrecker' in place with the hope of fighting the changes, or avoiding them, until they 'hopefully' go away. In some hospitals the competencies of Clinical Director applicants have been assessed before being appointed to directorates.

Tenure

A fixed tenure is desirable to obtain the benefit of fresh ideas from changing directors, but the Director should be sufficiently long in post to learn all facets of the job and to have time to put the acquired knowledge and skills into practice. The BMA/CCSC recommendation is for an initial tenure of 3–5 years and then renewable on an annual basis. There is evidence to suggest that some jobs are not fully learned for over three years, so it would be a mistake to move out before that time, or you may never fulfil your potential. It is equally true that a degree of challenge must exist in the job otherwise it becomes tedious and dull, something that may occur after ten or twelve years. A reasonable view is that about seven years is right, with the option to go earlier or be reselected on an annual basis after that. One of the problems about being too long in post is that it leads to a separation from colleagues, and credibility can be affected by being away too long. It should not be forgotten that jobs do change.

There should be a rolling tenure of appointments and retirements within the hospital to ensure that not all directors change at once. The appointment of a successor before a predecessor's retirement, with a 3–6-month overlap is highly desirable to allow the new incumbent to gradually and smoothly take over.

Succession

The succession should be planned and phased, and should not be by seniority. One of the problems for the first wave of clinical directors was the lack of management skills and training at, or prior to, appointment.

This should not be allowed to be an issue for successors. They need to be fully trained prior to taking up their appointment.

It might be worth a hospital or unit considering the possibility of input in the reselection or succession procedure, from other clinical directors.

Contract and job description

Ratification is a means of formalising the appointment. There should be a formal agreement with budgetary responsibility identified. Contracts are thought to be unnecessary by many hospitals. Participation in budget-setting excercise should be mandatory. Responsibility of different groups of staff within the directorate should be identified.

Any formal contract and job description should be separate from the clinical contract and job description and should specify that the job is part time and that the holder has continuing significant clinical commitments. Termination should be by three months notice by either the management body or the clinical director.

Support facilities

There should be sufficient support staff, with essentially a Business Manager and/or Nurse Manager together with secretarial support. Meetings of the directorate should be attended by all consultants in the directorate, Nurse Manager and or Business Manager where appropriate. The directorate meetings should be attended by all heads of department within the directorate. There should also be sufficient, accurate and timely information available to the Clinical Director, to enable him to carry out the work, readily and easily. And last, but by no means least, the Clinical Director needs support in meeting his remaining clinical commitments.

Time/remuneration

The Clinical Director should receive two or three sessions or notional half days per week or equivalent salary or any combination. Alternatively the extra sessions may be taken up by the other members of the directorate to cover dropped sessions, although this solution is rare as most colleagues are already over-extended. Some hospitals pay nothing but most pay on average about two sessions. Some, in the words of UGM/CEOs, are 'generous' or 'pay significant amounts' and others make the payments performance related. You can do one full time job well but two are impossible, and if you attempt it you end up doing neither job well. An additional problem is that the relationship between a UGM/CEO and his Clinical Directors is altered when they are paid to do the work than when they do it voluntarily, and there are many examples of this. But nothing is worse than the situation where some clinical directors are paid and others are not.

Appraisal

There have been suggestions that Clinical Directors should be appraised or subject to individual performance review (IPR) for their management role. Most UGM/CEOs consider this valueless, and indeed it is far from likely that consultants would participate.

Re-entry problems

This is an issue which has so far received little if any attention. What you will do with your time, how will you cope with relationships with your colleagues and re-integrate personally? You will have fallen behind in your practice, it will have been difficult to keep your skills up to date, your clinical sessions may well have been taken over by someone else, your perceptions and attitudes will have changed and you will no longer be sharing problems with those having a common interest. You will experience a sense of loss — loss of information, not knowing what is going on. You can help your successor by being supportive and it is important not to interfere or be critical of your successor unless asked for your opinion or advice.

There should be an absolute right to take up again any clinical sessions that were dropped to carry out the commitment as Clinical Director. You may feel that taking up your interest in teaching and research, which inevitably has suffered in order for you to undertake a management role will again fill your time and interest. It it important to be honest with yourself, you may in reality have been fed-up before and wanting something else, perhaps you were bored and found a new challenge. Whether found or wanted, management experience has given you a new perspective and, looking back, although it might have seemed rosy, you can never return, you should look forward and go onto something new using your new experience. But what is certain is that the vacuum will need to be filled and it is something to which you will need to address some thought.

Relationships with other managers

A number of authors have discussed doctor–manager relationships and without doubt they need to relate well with each other. How one achieves that state depends totally on the attitudes and styles of work of the individuals. You need the skill to work with the managers; the Clinical Director does not need to be a negotiator, finance officer, etc., those roles are played by the Business Manager and Nurse Managers as part of their role. The Clinical Director needs to understand and use the existing expertise of the team. This is where it takes time to learn the job and where development training is vital. It is helpful to have a dynamic environment, constant communication and smooth-working relationships based on an understanding of the role of colleagues. Responsibilities of other key staff within the directorate need to be defined and this will depend on the role model chosen locally. To whom should business and nurse managers report and how should they complement and support the work of the Clinical Director?

How can Clinical Directors exercise authority over and gain acceptance of clinical colleagues. Is it realistic to expect Clinical Directors to influence the behaviour and clinical work of colleagues and, if so, how might they achieve this? The BMA states that individual consultants have continuing responsibility for their patients, although that clinical freedom is subject to the limits of law, ethics, contracts, professional standards and resources so that Clinical Directors cannot commit colleagues to workloads or resource agreements, discipline or sanction colleagues, or override the clinical judgement of colleagues. He can, of course, negotiate on behalf of consultants agreed workloads and monitor that agreement. He can co-ordinate medical personnel matters, peer review and audit. None of this precludes any consultant continuing to talk directly to management about his practice.

Should any thought be given to the relationship between Clinical Director and Medical Director, Unit General Manager, the General Manager or the Chief Executive? Whatever the model, the relationship between the Clinical Director and Support Manager is the key to success or failure.

While Chapter 3, on management skills and Chapter 4 will expand on these relationships, the Clinical Director would probably look to these two colleagues for support, although it might well be a matter for individual choice as to where and how the various tasks of each role were split, depending on each individual's style, experience and personal preferences.

The business manager

The Clinical Director would look to the Business Manager for support by working with the Nurse Manager on day-to-day matters within the division and managing information within the directorate. Ensuring that the collection, collation, analysis and presentation of information (activity, financial and manpower) is carried out in the division to standards that are commensurate with those set for the acute unit as a whole. Coordinating the implementation of information systems within the directorate, and managing the necessary trouble-shooting.

Organising effective communication processes within the directorate

Specific tasks include facilitating the Director's requirements for management meetings concerning the creation, implementation and monitoring of directorate plans, and identifying cross-directorate issues that may arise.

Managing the non-medical and non-nursing staff accountable within the directorate

Specific tasks include managing annual/sick leave cover, economically whenever possible within existing staff levels.

Organising the distribution and allocation of work within the directorate

Specific tasks include creating and nurturing the correct level of clerical and secretarial support for the directorate as a whole. Agreeing annually with the Director a programme of tasks and objectives personal to the Business Manager for the financial year.

The nurse manager

The Clinical Director would look to the Nurse Manager for professional and managerial leadership within the directorate. To advise the Clinical Director on the best use of nursing staff to provide an efficient and cost effective service, ensuring the highest possible standards of care are provided.

There will, of necessity, be two lines of accountability:

(1) Director of Nursing Services on all professional issues.
(2) Clinical Director for all day-to-day issues in own unit.

The Clinical Director might reasonably look to the Nurse Manager for their authority to cover the:

- Line manager for nursing staff.
- Appointing staff up to registered nurse level as delegated by the Director of Nursing Services.
- Ensuring professional and managerial development of all nursing staff.
- Functional authority over non-nursing staff when patient care issues are involved.

The Clinical Director might expect the Nurse Manager to work closely with the Business Manager to ensure that the needs of the Division are met with particular emphasis towards:

- Ensuring highest possible environmental and care standards.
- Management of non-medical and non-nursing staff including advice on cost-effective cover for absence.
- Giving help regarding patient statistics and other relevant information needed to manage the directorate efficiently.
- Maintaining and improving communication within the directorate and ensuring the Director has all the information and assistance required to manage the unit effectively.

The medical director

The role is one of guaranteeing the delivery and development of quality medical services and ensuring this is the prime focus of the board of directors. Making sure that the board focus is maintained and for keeping the board informed.

The post of Medical Director may be advertised both inside and outside the hospital. It is not an elected or representational appointment. It may be a part-time advisory role or a full-time executive role. Some UGM/CEOs have shown considerable foresight in recognising the need for a full-time post and have advertised externally. The dilemma, as always, is that while a full-time post is more acceptable to the Board, it is generally not so acceptable to colleagues, who would prefer a Medical Director to retain some clinical commitment within the hospital or trust. However, a doctor in management is generally better than a doctor giving a medical opinion to management.

Applications should be invited for the position and a selection of candidates with the appropriate managerial competencies drawn up prior to interviews being held. Appointments should normally be made on a minimum three year contract. The dilemma facing suitable candidates is the same that has faced many doctors accepting a managerial role, and which has still not been satisfactorily resolved, is 'What next?' What do you do if you want to get out or are not able to do the job? Where do you go if your clinical job has gone? It is stated in the supplement to *Working for Patients* and the NHS and Community Care Act that the Medical Director is appointed by the Chairman, non-executive directors and the General Manager. Although it would be preferable that he has the confidence of the medical staff it is not a representative post but an executive one sharing corporate responsibility with the Board and responsible to the General Manager. Indeed, according to Johnston:[5]

> ... because no member of the Board is a representative it is one reason why chairmen of Medical Advisory Committee's should not be Medical Directors. Doctors will continue to require independent representative

machinery and it would be a confusion of roles to have both an executive and representative role.

The original working paper also suggests that it is possible to combine the role of Medical Director with clinical work. There is an expectation of more than regular attendance at board meetings to put forward the view point of doctors. In reality it requires a full time executive contribution to developing the services of a Trust, i.e. development and provision of comprehensive medical services and to advise on medical services and how they contribute to the aims and priorities of the Trust.

The Medical Director should be responsible for the discipline of all medical staff, and it is to him as representative of the Board that the Clinical Directors should report managerially.

In clinical research, the Medical Director should advise how research contributes to aims and priorities and develop policies which facilitate this. The Medical Director is also involved in medical education, for both undergraduates and postgraduates, including junior staff within the hospital and should ensure that the activities of all those involved in postgraduate education, including Clinical Tutors, Postgraduate Deans, Clinical Deans and the Royal Colleges are coordinated with the aims of the Trust. The Medical Director's role includes medical input for contracts for services and he should respond to the views of the main health authority and other districts concerning needs of the local community in relation to services provided. The Medical Director needs to be aware of the difference between what is needed and what is provided. A Trust with its own views will need to consult directors of public health about local needs and how they might best be met and the Medical Director will need to resolve conflicting views. The Medical Director will also need to review outcomes of medical services and impact on population needs locally and to facilitate and coordinate the identification of new and changing areas of need relevant to the services offered by the Trust. The Medical Director must also represent the views of the Trust to potential purchasers, together with a public relations profile for dealing with areas of potential conflict. In these matters only a doctor can lead the input.

Good relations with general practitioners are very important to a Trust. The Medical Director should facilitate and encourage this process, including ensuring that opportunities for professional development are accessible. It is also part of the Medical Director's role to go out into the community and tout for business.

Viewing the development of medicine within a hospital within a 5–10 year perspective is not an easy role to fill in the NHS. Development of medicine has not been a priority and hospitals usually lack an identifiable person responsible for that position, let alone someone with the vision to carry it out.

The current approach to development is largely *ad hoc* and will not serve medicine or hospitals well in the future. There is a need for clarity about the role and direction of medicine as a whole within the hospital, ensuring that in conjunction and cooperation with others, doctors become appropriately trained, are properly involved in managing hospitals and the setting of its future direction. Trusts which fail do to this will fail to thrive. The changes which we have seen in the hospital service recently are as nothing to the changes that will take place in the future. It is a downsizing industry and there is a lot of demarketing going on throughout the world. There is no training available for this challenging role, training in public health medicine is not appropriate, although some of the epidemiological skills would be appropriate.

References

1. Nichol D. Report in *BMA News Review*, 1991; **17**(1): 10.
2. Chantler C. Personal Communication, 1991.
3. CCSC. *Guidance on Clinical Directorates*. London: BMA, 1990.
4. Kennedy P. A better way than Clinical Directorates. Health Services Management. October 1990, pp. 211–15.
5. Johnston IH. What will the Medical Director do? *BMJ* 1991; **302**: 280-81.

3

Managing as clinical director
Tony White

Introduction	40
Mandate for a clinical director	41
Tribalism	45
Management skills	46
Corporate management	53
Communication	53
Characteristics of success	54
Technical and managerial competencies	55
A breath of hope	56
Some dilemmas	57
References	58

Nam et ipsa scientia potestas est. (Bacon)

Introduction

Making decisions in a sensible way is a human function that, when it involves other people, requires management skills. Taking action is also a human activity but, again, when it involves other people it can be more efficient and more effective if management skills are used. There is unfortunately no universally accepted definition of the management process. It is certainly not an exact science and for that reason often does not appeal to a doctor's basic instincts and training. The search for universal principles is frustrating. Sweeping generalised principles cannot be developed because what works in one setting may not work in another. Studying success and failure, i.e. what managers do and why, is often more useful.

Management is an art with a science base. It has generally to be learnt by experience. Being complex it is interpreted differently by different people. The manager is concerned with setting goals and achieving objectives. The effective manager is many things, a historian learning from past success and failures; a psychologist who must understand the way people act in, and react to, group situations and an innovator who can develop new ways to achieve desired objectives and apply them in an appropriate manner.

Health service organisations are unique in a number of ways and management study and practice has to take account of that uniqueness. These special features include the absolute necessity for high quality of work, the involvement of high technology, the utilisation of a wide range of human resources and the coexistence of automated and manual work methods, etc.

The carrot of the latest reforms is that with less bureaucracy quicker decisions should be made. Significant service developments and initiatives should be achievable in a surprisingly short time. Long-standing minor but persistent irritations should be swept aside; long-term conflicts of interest are brought to a head and resolved.

Mandate for a clinical director

There are three vital requirements for a successful Clinical Director: (i) responsibility; (ii) accountability; and (iii) authority. Each requires defining and in particular the issue and problems of authority, which some managers find a difficult concept to grasp, should be made clear. To fulfil duties what degree of authority needs to be vested in a Clinical Director? The division of responsibilities must be clear, without this clarity, accountability will be clouded.

Responsibility

Definition of roles and responsibilities is relatively easy to agree. But responsibilities to whom? The British Medical Association CCSC[1] states unequivocally:

> Clinical directors should be managerially responsible to the management body.

But also to be considered are the faculty, the unit, the directorate itself and the speciality. And responsibilities for what? It is important that this is identified and agreed. Again the BMA/CCSC view is that budgetary authority should be for staff, i.e. medical, nursing, secretarial, administrative and other professionals and non-staff, i.e. medical and surgical equipment and drugs, ward supplies, etc. But the reality depends on the role model chosen and local decisions and agreements.

There is also the question of loyalty and responsibility. Indeed, how different are they? Responsibility implies being accountable. Does loyalty mean adopting hospital policy and accepting the constraints of management like mindless conformity or should it mean doing what one thinks is best for the organisation in the long term? Moss Kanter[2] offers a theory in post-entrepreneurial organisations of the prospect of a different definition of loyalty:

> ... there is often no such thing as a 'chain' of command, and people work under different leadership for different purposes ... There is encouragement for people to test limits, challenge traditions and move in new directions ... Decentralization of decision making responsibility puts more power in the hands of people at lower levels to make decisions and exercise judgement ... Professionalism ... transcends the organization ... Post entrepreneurial organizations produce so much change that they cannot offer the same incentive for unquestioning obedience.

Accountability

Together with responsibility and authority comes accountability. Better financial information is sought, budgetary awareness is heightened and performance is more closely monitored. The new responsibility requires accountability. Accountability not only for the level and quality of service but also to the production of data that demonstrates and confirms this. Clinical audit and quality management suddenly become relevant and continuous multidisciplinary activities. Because standards are largely determined by doctors themselves, they are understood, owned, relevant and achievable.

Authority

Managers and Chief Executives balk at the thought of giving doctors more authority. They talk of authority in terms of 'hire and fire' and do not see the wider context. How can a Clinical Director exercise authority over, and gain the acceptance of, clinical colleagues? Is it realistic to expect the Clinical Director to influence the behaviour and clinical work of colleagues and, if so, how might they achieve this? The BMA[1] argues that individual consultants still have continuing responsibility for their patients, although that clinical freedom is subject to the limits of law, ethics, contracts, professional standards and resources so that a Clinical Director could not commit colleagues to workload or resource agreements, discipline or sanction colleagues, or override colleagues' clinical judgement. The Clinical Director can, of course negotiate, on behalf of consultants, agreed workloads and monitor that agreement and can coordinate medical personnel matters, peer review and audit. None of this precludes any consultant talking directly to management about their practice.

Talk to doctors about authority and they mostly think of authority to encourage their colleagues in a directorate to do what is asked and required of them. But authority for a Clinical Director is more complex than that. He does not need what the managers talk of to do his job, and he does not need to be given authority over his colleagues, who have already put themselves into a quasisubmissive role by accepting him as their Director or agreeing to his acting in that role.

So both doctors and UGM/CEOs seem to exhibit considerable confusion over the authority required by a Clinical Director to undertake his role. There are three definitions of authority according to the Oxford English Dictionary:

(1) Power or right to enforce obedience, which the Clinical Director does not have and which is not relevant to the role of a Clinical Director or to this discussion.
(2) Delegated power, which comes from the CEO and Management Board and which is very reluctantly given, as decentralisation and devolution of authority is a generic problem.
(3) Personal influence, especially over opinion, which is earned and comes from respect. Despite the lack of line authority some clinicians have considerable success in managing their colleagues. The processes that the successful ones use are based on representing, involving, consulting, trading, using personal power, the skilful use of information and often the ability to fit their work into a broader picture are all discussed later (see p. 46). There are three approaches with personal influence. The so-called good, the bad and the logical:
    (a) Christian (good). Dealing with brotherhood, love and trust.
    (b) Machiavellian (bad). Wheeling and dealing, scheming, fixing, politicking and knife in the back.
    (c) Logical. Using facts, logic and sensible and rational argument.

A Clinical Director will need to have all these cards in his hand, and he may well have a natural tendency to play one more than another, but what is important is knowing when each should be played, i.e. *savoir-faire*.

However, not all the tasks required of a Clinical Director are within the sphere of personal influence and it is here that problems arise, when Management Boards expect results which clinical directors cannot deliver. When the Board thinks that what is required is more accountability. If you ask a surgeon to arrange for his directorate to carry out a contract to perform a number of elective operations, he may persuade his colleagues and they may be very willing to do that. However, there are likely to be a number of stumbling blocks. The directorate may not have the time within the week, they may not have the beds, they may not have theatre time, there may be no available anaesthetic time and so on. Unless the Clinical Director has authority to organise these issues, no matter how willing the directorate to carry it out it cannot be done.

It is all very well for the Hospital or Management Board to say that there is spare capacity in the system, there may be beds closed, but the Clinical

Director may have no authority to open them. There may be spare capacity in the theatres, but if the theatres are working to a budget and have overspent already, the Surgical Director has no authority to change that. There could be no available anaesthetist, and again the Surgical Director has no authority to alter this situation.

In outpatients there may be too many people waiting in one town while the clinics for another country hospital may have too many clinics available for the existing demand. The Clinical Director in many hospitals has no authority to change that service provision without extensive consultation with management. There may also be more than one Trust involved and the local general practitioners may invariably refuse a reduction in services. In other words, the Clinical Director has no authority to change the service in response to need.

The number of outpatient clinics being carried out within a faculty or speciality may be greater than the number of operating sessions that can service the number of patients requiring operation. The result may be an increasing surgical waiting list. This will need a change in the balance between outpatient clinics and operating lists. But the Clinical Director may have no authority to change that.

It is thus possible to see that to give a Clinical Director the responsibility for a job, no matter how accountable you make him, and there is a lot of talk of accountability, you do need, in addition, to give him the authority to 'get the job done'. Unfortunately this vital ingredient is missing in some units, and it is here that the crux of many a problem lies because it comes back to the way in which the system works.

The system by which hospitals work has been changed. No longer are they given a fixed budget and expected to provide a limitless service, they now 'earn' money from contracts for treating a certain number of patients. Part of those contracts are to provide an emergency service, but a significant part of their contracts from both the main district provider(s) and general practitioner fundholders is to undertake non-emergency work in line with the service agreements. The problem, however, often arises that this new system has not filtered down to directorate and departmental level. There they are still having to work with fixed budgets, often having to carry items for which no budget has been devolved, such as travelling expenses, stationery, MRI scans, ward equipment, interview expenses, advertising, etc. and supply a service at a level that is negotiated and agreed with the purchasers at a level above the directorate, and into which the directorate has no input.

The patients may bring money to the hospital but they bring no money directly to the departments or directorates. This especially affects the so-called service departments. Anaesthetics and theatres have to provide a service on a fixed budget. They cannot do extra lists because that will be a drain on their budget. Radiology and pathology are subject to the vagaries of demand from the clinical specialities on a fixed budget. The only way that radiology can limit demand to remain within budget is to increase the waiting time for radiological investigations.

The problem appears to be that, at hospital level the situation may have changed but at directorate level the old system may remain. The system was condemned by management as the reason for past failures but has been pushed down the line now that a new era at their level has arrived. The problem is that it is failing at that point for the very same reasons that it was unsuccessful before.

What is required is a system of internal marketing at directorate level where money follows the patient. When management or fund holders want extra patients treated, or extra contractual referrals are referred to the directorate, the Clinical Director should be able to use the money that comes with those patients to purchase the necessary services from theatre, anaesthetic department, radiology or pathology. These service specialities would then in turn be able to fund whatever degree of service was required of them. This system is already operating as a general principle in some places, with some departments able to operate such a system more easily than others. Until it becomes a universal practice there will inevitably be cases where fund holders have the patients requiring treatment, the money to pay for that treatment and clinical directors prepared to carry out that treatment, but no authority to get it done.

One further issue of authority is the control that Chief Executives and General Managers have over Clinical Directors. Most Chief Executives and managers express the opinion that they would take an extremely serious view of failure of a Clinical Director to achieve the objectives of matching activity to resource. The ultimate sanction of course must be that he would be removed from the role. There would be a twin loss, of the remuneration for the work, although this is hardly commensurate with the work involved, and the loss of face if this was the result of incompetence. However, a few hospitals pay nothing in time or money for their Clinical Directors, but rely on goodwill, dedication and loyalty and it seems probable that removal from office of one of these is likely to be followed by a sigh of relief that the responsibility is no longer carried. It is a role that, without the right support and authority, is likely to achieve so much less than would be possible with the right understanding of the set up.

Tribalism

There is the strongest identification with groupings of patients rather than a profession. In other words doctors identify with their department and speciality. They feel only secondary loyalty to the hospital. If you ask a consultant what he does for a living they invariably say 'I'm a surgeon' 'I'm a doctor'. They never say 'I work for X hospital'.

Similarly staff feel more responsible for standards of care throughout the directorate rather than exclusively those of their own profession. Tribalism is confronted in the face of more explicit shared responsibility. There is more relevant participation in the running and development of the service. There is commitment to corporate management. This entails a consciously managed process that must be seen to erode traditional tribalism. It encapsulates the difficult shift from fighting one's corner to fighting for someone else's.

Management skills

Just as there are different levels of commitment (see Chapter 6) so there are different levels of commitment by consultants to management:

(1) Leaders. Those who understand and are committed to the contribution of management to patient care.
(2) Helpers. Those who see it in their interest to become involved in what they see as management.
(3) Followers. Those who can detect a shift in the balance of power and realise that it is in their interest to become involved in management 'to be on the winning side'.
(4) Opposers. Those who are oblivious to management and who are here 'to treat patients'

People can of course vary according to mood and circumstance but on the whole it is an evolutionary process to proceed up the scale; except for those dinosaurs stuck at the bottom. Some of those at level 2 can also be dangerous, being there for the wrong reasons and seeing no reason or value to aim for level 1.

It might be helpful at this point to consider what it is that management hopes, demands and expects from doctors involved in management. This has been summarised as follows:

- Demand integrity.
- Hope for the best.
- Expect considerable variation in performance.

What they get is varying levels of competence at the job:

- Natural ability. Although that ability may be doubted.
- Trainability. Those who could do a reasonable job when trained.
- Unbelievability. Those who think they can do it but never learn how.

Mintzberg's classic research[3] on managers' ten roles suggests that they are under constant pressure to acquire and disseminate information, to develop strategies without time for analysis, to influence the behaviour of others without being dictatorial, to react sensibly to external initiatives without creating an impression of weakness, all of which require a manager to develop a network of relationships that depend critically on the art of communication.

Mintzberg[3] refers to management roles as being:

(1) Interpersonal roles:
   (a) Leader.
   (b) Figurehead.
   (c) Liaison.
(2) Informational roles:
   (a) Monitor.
   (b) Disseminator.
   (c) Spokesman.
(3) Decisional roles:
   (a) Planner.
   (d) Disturbance handler.
   (c) Resource allocator.
   (d) Entrepreneur.

Handy[4] puts the three roles more colloquially as:

- Leading.
- Administrating.
- Fixing.

Interpersonal roles

These are the leadership roles. Doctors have difficulty with leadership roles. Consultants especially see themselves as autonomous. Primarily a responsibility of the Clinical Director rather than the Business Manager or Nurse Manager. A Clinical Director is the leader of the directorate and has to take full responsibility for devolved operational management. He needs to be creative, ambitious and a leader. High motivation and enthusiasm is required to set the style and pace of the directorate and its management. The Clinical Director is accountable, leads the decision-making process but always ensures that they are empowering others to make decisions — an enabler and facilitator, but always pursuing excellence.

The Clinical Director must have a clear responsibility to determine directorate organisational behaviour and standards. He is a powerful voice in what the directorate does and how it does it. The Clinical Director is responsible for issues of audit, review and evaluation, staff development and appraisal, and also determines whether the directorate has a high or low profile.

The management style of the directorate is set by the manner, behaviour and actions of the Clinical Director. Incumbent on the Clinical Director is also the responsibility of being role model for many others — a major responsibility is to provide an effective and successful role model that other staff wish to emulate.

Leadership is the ability to inspire and influence others to work towards the attainment of your objectives and goals. This is necessary because of the task of getting work done by and through others. It is therefore necessary to choose a method of leadership most appropriate to the situation. People need to be motivated, and the best motivator is challenge, which allows a feeling of achievement, although it must be followed by recognition.

A Clinical Director needs strong leadership qualities, establishing a directorate from scratch, and then running with the help of Business and Nurse Managers the management process within the directorate. The Clinical Director needs to create a team spirit not only amongst his immediate team (the Business and Nurse Managers), but amongst all the members of the directorate.

The Clinical Director needs to be able to recognise good ideas, analyse problems, make decisions, accept responsibility for those decisions and be accountable for those decisions. The scope for leadership and flair is boundless. There should be established a well defined directorate leadership, which acknowledges respects, values and uses to full capacity professional differences, similarities and contributions. It will be necessary to invest people with a sense of identity in the directorate, helping medical staff to adapt and contribute fully to the new structure and process.

Delegation of leadership means people are more in touch with leadership. Taking responsibility and accountability, and provided with the appropriate authority, the doctor is able to be part of the corporate hospital management. The Clinical Director will need to establish clear complementary and effective roles and responsibilities with other managers in the directorate, reviewing and updating them as skills develop. He will also need to promote clear informative relevant and comprehensive communication; promote review and evaluation; represent directorate business at corporate level; develop corporate responsibility and skills in himself and within the directorate; demonstrate motivation, commitment and enthusiasm for the management process.

Informational roles

Many of these are the administrating roles that are primarily a responsibility of the Business Manager.

Information management

Information management has three main roles: (i) to monitor; (ii) to use this information for problem analysis and solving; and (iii) to disseminate information, i.e. to effectively inform colleagues and all other members of the directorate, following management decisions. It is important that colleagues and management understand the importance of accurate and timely information. It is equally important that information available within the directorate *is* accurate and timely. It may be necessary to review and refine available information.

Coordinator

To act as spokesperson, coordinating medical input into management information. The Clinical Director is the main link with other directorates and the hospital management system. The sentiment and content of messages going out of the directorate must be realistic and accurate, and the calibre of incoming messages must be loud and clear. Information and communication skills and networks are important and the Clinical Director can use his considerable influence to ensure that these are good.

Consulting with others and taking advice is about building strength, not admitting weakness. Many Clinical Directors feel particular vulnerability in expressing a need for guidance, support and learning for themselves.[5] This can hinder personal, professional and directorate growth and is not a good example to colleagues. Directorate culture must be to support all staff development within the group. External resources provided by other sites, academic institutions and organisations should be fully utilised for this purpose.

Monitor

Responsibility for the quality management of the organisation is brought closer to the patient with an impact on the quality of service delivered. Quality is understood and embraced as an integral part of everything everyone does. The patient should be the primary focus of the efforts by all staff. Success is defined in terms of quality. Achievement should be recognised, broadcast through the organisation and rewarded. People then begin to see the wider impact of their own particular and unique contribution to the success of the whole.

Part of the current role of a Clinical Director is to ensure that effective audit mechanisms are in place, that reporting of audit is occurring and that improvements and changes are monitored (this role is dealt with more fully in Chapter 5). The managing of a balance between quality, quantity and cost, particularly quality in contracts, will probably be part of the Clinical Director's future role.

Decision-making role

This is the fixing role. Although the role is less clear-cut, divisions between

Clinical Director and Business Manager, planning and resource allocation being more predictable, it is primarily a role for the Clinical Director.

Vision is required, as is a capacity for growth and learning. An outlook and an approach to do with evolution and optimism replaces doubts and anxieties about change. Implementation of change is seen as a developing and flexible process in pursuit of a much-desired goal. Feeling more positive about change, people develop an ability to live comfortably with uncertainty and apparent contradictions. The Clinical Director is responsible for responsiveness to change using his discretion to act but always in a responsible manner. Uncertainty becomes the expected way of life rather than an occasional threat. There is a greater emphasis on solutions and opportunities than on problems, on creativity rather than ritualism and control.

The greater flexibility demanded of staff promotes greater learning. Innovatory practice is valued and rewarded. Implicit here is the expectation and acceptance of making mistakes, learning from them and modifying practice accordingly. The requisite risk taking and its consequences are supported.

Dealing with conflicts is often a problem. Doctors have problems with negotiation and conflict and have difficulty in believing in teams. They see committees as looking after territory rather than the whole organisation. They also have difficulty thinking strategically. You may need to make a stand on critical issues. Problems must be faced openly and honestly. This relates very much to working with people, being aware of others and seeing things from another person's perspective, to read or interpret other people's thoughts and feeling even when they are not obvious, developing others and working as a team.

Change management

Change is never easy. People resist it for many reasons — insecurity, social and economic loss. It is an issue addressed in more depth in Chapter 7. Kaluzny and Hernandez[6] identify three types of change as a function of whether ends or means or both are involved:

(1) Technical change involving modification is the means by which activities of the organisation are carried out. There is no change of goal.
(2) Transition change, in organisational goals but not in the means of achieving those goals.
(3) Transformation, the most extreme form of change, where there is change in the goals and the means to achieve them.

The process of change is a step-by-step process. First there is the recognition of the need for change. Then the problem needs to be identified, and with it alternatives methods and strategies of dealing with it. The selected method and strategy then needs to be chosen and the change implemented. And finally an evaluation of the success or otherwise of the change needs to be made. This is important, as the Clinical Director has a responsibility to utilise resources to the optimum. It can also prove whether the change was wise or whether further change is necessary.

Negotiator

An increasingly important role for the Clinical Director is in influencing the activity of the directorate in the analysis and identification of opportunities and the encouragement and development of awareness of business planning in colleagues. The Clinical Director needs to:

- Lead the discussion to analyse issues and make decisions.
- Agree realistic and achievable activity levels through contracts.
- Ensure that the business plan and the subsequent monitoring activity are acceptable to colleagues.
- Hold meetings and discussions with medical and other staff before agreeing work patterns and targets within the directorate.
- Monitor progress through the year.

It will also be necessary to be responsible for analysing progress against the plan and expenditure against budget at monthly intervals and taking appropriate action if the results are not matching targets.

Although in essence a work agreement, within the available resources, preparing and arguing a business plan should include an opportunity to bid for a budget increase for the directorate each year. This will require good information in support of the case and it is the role of the business and nurse managers to ensure that such information is timely, reliable and available to the Clinical Director.

Financial management

The Clinical Director needs to get 'hands on control of the money' if he is to play any meaningful role in management. He needs to ask who really makes the decisions and how. As an example, take a recent piece of equipment and find out who really made the decision to buy it. If you are to be part of the management structure you, as Clinical Director have to be part of that process. You also need to be informed of the total budget position of the hospital, your budget in isolation is meaningless. You need to be honest with yourself as well as the other members of the management team about your financial position. It is no good saying 'you are doing fine' when you are overspent; it is no good saying 'it is only a little overspend' and it is no good retreating into the defence of it, if it is not really an overspend but an underfund. Those days are gone. You are the managers now. It is important that Clinical Directors, Business Managers and Nurse Managers meet with the Financial Director and for each director to state his financial position month by month. The Clinical Director must direct and support negotiation and management of soundly based budgets.

Depending on the role model chosen and local decisions, the Clinical Director may accept financial responsibility for provision of defined services, but only after his personal involvement in negotiation. He should certainly encourage financial awareness among colleagues and participate in financial review of the directorate. He should advise on clinical priorities to ensure the directorate stays within budget but he needs mechanisms in place to monitor financial performance so that corrective action can be taken as appropriate, operating within the statute, rules and conventions of the Trusts or self-governing unit. You will require a better understanding of budgets, financial planning, understanding and managing contracts, costing and pricing of services and the interpretation of financial information, possibly even encouraging colleagues to identify income-generating opportunities.

The Clinical Director's budget is likely to be made up of about 80 per cent pay, i.e. staff salaries, and 20 per cent buy, i.e. materials. If there is an overspend, the Clinical Director needs to agree a figure for this with the Finance Officer and have this in writing over his signature. Computer printouts have a habit of changing or being changed. By this stage the Clinical Director may well find that the Finance Department has already discovered some errors and the overspend is not as bad as was originally stated. He should then identify every single item, missing out nothing. He may need to take some time to discover these but nothing should be accepted on trust. All requisitions should go through the Clinical Director and be signed only by him. He should look at maintenance contracts for items of equipment that might long since have been abandoned and look at estate items closely; always being on the look-out for double invoices.

On the income side, the Clinical Director should make sure that he has at least one line set aside for income. Directorates need to retain at least 50 per cent of savings made, and a large proportion of the income generated. The remainder will go in a form of 'corporate tax'. If all the money, or an unreasonable amount of money, goes to the corporate level then people will not play the game for very long. The rules need to be set out beforehand. Management retains its power by not setting out the rules beforehand.

Contracts management

The Clinical Director should direct and support the Service Manager in negotiating and agreeing achievable and realistic contracts, and should then manage and control clinical activity to ensure that the contract requirements are met. The Clinical and Service Managers should ensure that mechanisms exist to manage and control activity levels to contract requirements and coordinate medical input into establishment and maintenance of accurate information on activity and costs.

Corporate management

There are three main areas to this. First, a two-way process of communication between the directorate and management, providing in one direction a clinical perspective to corporate management. Offering clinical colleagues the opportunity to debate and influence management policy decisions, it also enables the views of clinical colleagues to be represented to management. In the other direction it ensure colleagues are informed of management decisions.

The Clinical Director should be participating in hospital strategy and policy formulation contributing to effective decision-making within the hospital, for which he should recognise corporate responsibility. While at the same time he should be setting and monitoring directorate objectives and ensuring they are consistent with the overall objectives of the hospital.

Communication

Everyone acknowledges the benefits of comprehensive, open and effective channels of communication. The most favoured method is the cascade principle. It is the Clinical Director's responsibility to ensure that information is passed throughout the directorate. Lateral communication below Board level is important.

Does your hospital listen as readily as it talks. Do you as Clinical Director have an adequate voice in the corporate decision-making process? If not, this needs to be changed. How many Clinical Directors are the most effective number for effective corporate management and decision-making. The BMA/CCSC[1] emphasise:

> ... the requirement that the management body is not too large to be unworkable.

Maybe your hospital is still running with the old organisational structure while paying lip service to the new? Is the management structure of your hospital the most effective forum for corporate decision-making?

Communication should involve three factors: (i) talking; (ii) debating; and (iii) listening. Effective communication to staff is vital. You need to express what has to be done, with confident and clear instructions. How it is to be done? By whom it is to be done and when it is to be done? In a highly complex organisation like a hospital this is a multidirectional process. It needs to reach superiors, subordinates and peers.

Ensuring good communication within the directorate and between other directorates is a process of people relating to one another. Listening, demonstrating empathy, checking, understanding and providing feedback. It is vital to prevent and diffuse potential areas of difficulty with staff and colleagues.

You need, as Clinical Director, to relate in groups or one-to-one situations, to be able to lead discussions and meetings competently (this is discussed more fully in Chapters 7 and 8), to prepare and present arguments based on facts and to resolve claims on competing resources.

Characteristics of success

Certain characteristics distinguish the most successful doctors in management. Management research has increasingly focused on observing what managers do rather than on what they should do. Boyatzis[7] tried to identify the characteristics of excellent performance. He concluded that a job is performed most effectively when three elements are congruent: (i) the job demands; (ii) the organisational environment; and (iii) the competence of the job holder. Competence for the task involves personality, values, motives, attitudes and behaviour, as well as skills and knowledge.

There are technical skills associated with the particular profession or speciality. These are basic skills and knowledge possessed in order to achieve success in a professional role. There are basic managerial skills appropriate to any managerial situation, for example leadership, people management, social skills, supervision and working with others. There are also special skills that distinguish the excellent manager. Some are personality features, which cannot easily be learned. Others are more amenable to change or can be taught and learned. Making these distinguishing skills explicit can be an agent for change.

The Clinical Director needs to demonstrate enterprise and initiative, be decisive, get things done and try to improve the service. Making things happen involves setting goals and achieving them, shaping the direction of the directorate, providing a clear focus for the work of other members of the directorate, setting standards, delegating tasks to get the work done and organising resources to achieve objectives.

Improving the service means doing things better, either faster with fewer resources to a higher level of quality, or achieving improvements in both effectiveness of the service and in the efficient use of available resources. The Clinical Director should also look for and recognise inefficiencies in current practice and generally look for ways of improving how things are done.

The Clinical Director needs to be able to think conceptually, analytically and strategically; breaking issues into component parts to connect into a coherent whole, in other words to think about the totality of the problem. Can you identify the key factors in a complex situation and question the basic premises and assumptions, conceptualise frameworks to relate issues to the broader picture and makes connections between different parts, develop and use clear criteria for evaluating options, anticipate problems and develop contingency plans.?

Can you envision a long-term picture of the future of your speciality and directorate based on an analysis of its purpose and the future environment in which it will be working and develop strategic goals and demonstrate learning and change in behaviour as a result of experience?

Influencing strategy, persuading, trading and negotiating and planning intervention are analytical thinking skills that feature highly amongst doctors due to their scientific training. However, one method of influencing could be inadequate. You may need to utilise several approaches including: Identifying the key people who need convincing. Tailoring influencing strategy to the concerns of key individuals/groups. Using networks to gain support. Using personal relationships to short-cut the bureaucracy. Lobbying influential people in advance of formal meetings. Keeping key people informed on issues. Elaborating logical arguments to influence. Presenting or organising data in order to influence. Using cost benefit argument to influence. Appealing to the greater good of the hospital or community.

Turrill et al.[8] in a study of the excellent doctors in management referred to these types of characteristics as competencies and found that:

> These occur with sufficient frequency to suggest that they are too important to the role to be ignored. They may be termed the threshold variables in that they suggest the minimum conditions for fully acceptable performance.

Achieving
  Demonstrating enterprise and initiative

Thinking
  Thinking analytically

Influencing
  Influencing strategically
  Persuading rationally.

Technical and managerial competencies

Technical and managerial competencies are eminently trainable. These primary behaviours and the distinguishing competencies may be improved, but only within the limits of the individual's innate potential. They result from the underlying characteristics of their personality and are the building blocks for managerial process competencies such as team leadership, negotiation, etc.

Possession of these attributes alone is insufficient. Success is unlikely unless the clinical management teams acquire the full range of managerial process skills and the unit has management systems in place that are appropriate to the new arrangements. This is unfortunately not always the case.

Some doctors display a somewhat restricted range of managerial skills. Unless the managerial systems are in place and the clinicians have the managerial skills to use them, the Clinical Directorate initiative will be frustrated. The overall climate is set by the Chief Executive, and the managerial systems provide the arena in which successful Clinical Directors can display their primary behaviour using their managerial process skills.

Some outstanding doctors have a very clear view of their own role and the part they have to play within the overall process. Their common view is that management is about trying to do more with less, pursuing excellence with limited resources. For many reasons not least of which is the time they have available, it may be that they are better placed to do this if they act as transforming leaders rather than transactional managers. Their choice of role will have a significant effect upon the process competencies required by them and their immediate team.

The operation of a successful directorate often rests heavily upon a team approach and the Nurse Manager and Business Manager often play a significant part. The capability of this group of staff is critical. A senior nurse or technician who may have been successful in different situations will not necessarily have the correct distinguishing competencies for this new role. It may well be that the key position for the lead clinician is leadership. If so, it is important that the Business Manager and Nurse Manager develop their managerial process skills to support the leader.

A breath of hope

Stewart[9] describes the job of managers as made up of constraints, choices and demands:

- Demands include:
    - Minimum criteria of performance.
    - Procedures that cannot be ignored.
- Constraints include:
    - Resource limitations.
    - Physical location.

- Attitudes and expectations of others.
- Choices include:
  - How work is done.
  - What work is done.

The individual has some discretion in shifting the balance (by pushing back constraints, for example). (Stewart[9])

Some dilemmas

The dilemmas that need to be taken into account as directors develop their role are the professional and organisational tensions. No longer are they just a clinician and equal member of the peer group, but leading a management team while at the same time engaged in clinical activity. Respected clinician maybe, but now a complete beginner in management skills. As Clinical Director you are fully responsible for performance and accountable for that. But at the same time your involvement with day-to-day decision-making is part-time, may be peripheral and delegated. With perhaps a passion for one's own speciality the problem is how to make room for understanding and being concerned about others. Having to learn to represent corporate directorate interests, and to put these aside for unit corporacy.

The Clinical Director must not give in to the pressure of being all things to all people. He should not try to be informed of all directorate business and be party to all decisions taken, as this is not only unrealistic but intrusive upon others and obstructive to their proper functioning. The Clinical Director has, too, to do his fair share of devolving authority and decision-making. All staff need to have confidence in their Director's common interest in all and the holistic approach to the directorate. This key player must be trusted to sacrifice self-interest for corporate well-being and development. Staff motivation and goodwill will be lost if they see any sign of their Director seeking personal or professional power rather than equity and accountabilty.

A further dilemma is how to cope with a UGM/CEO who appears to lack the ability or confidence to devolve. Poor organisations do not always think through what they need or ought to do and, if they do, they do it poorly. They do not plan the implications of change for the staff, do not train the staff for the change and lack the right sort of communication. A bureaucracy is very difficult to dismantle, but is it possible to rehabilitate a UGM/CEO in difficulties, with the changes taking place in the health service. When did your UGM/CEO last come and see you operate, or see any operation? To be involved and understand the organisation is a fundamental requirement. Hospitals will look radically different in 20 years time. It is a downsizing industry with demarketing going on all over the world. An increasingly ageing population will not alter that basic concept. Primary care is taking on more and more of the functions, hitherto the sole territory of hospitals. Day surgery is increasing but at present is only a fraction of what it will be achieving in the future. Yet in spite of these signs most US and UK Trust hospitals say in their plans that they envisage continuous expansion over the next few years.

Doctors are able to provide long-term continuity, to balance the constant change of managers every few years. This leads to managers thinking shorter-term than previously. In fact, many poor managers are supported by the medical staff. Neither the manager or the medical staff may even be aware of this fact, such is the loose management structure that some hospitals have. It is sometimes necessary to understand the UGM/CEO perspectives. He is in a career too. You need to understand where he has came from. You need to understand the influence of his mentors. Who are his referees and what is their background. Even if he came from outside the health service there is likely to be a role model contact within the service. There is usually a supporting network. You need to study the whole person, his career to date and his mentors, then you will understand him as an individual.

References

1. British Medical Association. *CCSC guidance on clinical directorates*. London: BMA, 1990.
2. Moss Kanter R. *When giants learn to dance*. London: Unwin Hyman, 1989, p. 338.
3. Mintzberg H. *The nature of managerial work*. New York: Harper and Row, 1973.
4. Handy C. *Understanding organizations*, 3rd ed. London: Penguin, 1985.
5. Wraith M, Casey A. *Implementing clinically based management. Getting organisational change underway*. In conjunction with The Resource Management Unit of the NHS Management Executive, 1992.
6. Kaluzny AD, Hernandez SR. Organizational change and innovation. In: Shortell SM, Kaluzney AD, eds. *Health care management: a text in organization theory and behaviour*, 2nd ed. New York: Wiley, 1988, pp. 380-81.
7. Boyatzis RE. *The Competent Manager*, New York: Wiley, 1982.
8. Turrill T, Wilson D, Young K. *The characteristics of excellent doctors in management*. NHS Management Executive, Resource Management Unit, 1991.
9. Stewart R. *Choices for the Manager*. New York: McGraw-Hill, 1982.

4

Nursing management issues
Anne Cooley

Introduction	59
The role	62
The relationships	64
The responsibilities	67
Manpower	71
Developments in nursing	73
Dilemmas	76
References	77

Tempora mutantur et nos mutamur in illis.
(Emperor Lothar I)

Introduction

No discussion concerned with the role of doctors in management would be complete without some consideration of the part played by nurses within this scenario. Ever since the early nineteenth century, when hospitals began to be peopled by medical students and junior medical attendants, a nurse has been at hand to help:

> ... with dressings, but only in fetching tins of warm water with which the doctor cleaned the wound. She was allowed to apply either a bread or linseed poultice but as soon as dressings or lotion or lint were ordered the pupil will take charge. If the state of the patient was such that someone had to sit up all night it was not the nurse who did so but a pupil. The pupils were apprentices of the physician, surgeon or apothecary.[1]

The matron of this time held responsibility for the whole internal working of the hospital. For cleaning, for the cooking and serving of food, for the laundry service, the provision of linen and the stores. In addition, and considerably lower on the list, was the organisation of nursing. Primarily nursing was in the hands of doctors and the lay administration. Under the influence of Florence Nightingale, however, in the mid-nineteenth century the matron became supreme in matters of nursing. She held that whilst nurses must follow doctors' instructions without question in the administration of treatment, nursing must increasingly develop as a unique skill; one managed by and accountable to a head nurse or matron. A further considerable influence was exerted on nursing by both the Church and the Army. Under this influence nursing developed as a hierarchical and authoritarian profession. Symbols of this influence remain in the title Sister, the uniform and the terms used, i.e. 'nursing duties'. Towards the twenty-first century, nursing as a profession is striving to shake off some of its earlier influences and to develop independent nurse practitioners who are individually accountable for their practice. Not reliant on a decision-making hierarchy bound by protocol and dictat, but rather recognising and grasping the authority and the responsibility to act at patient care level. Nurses will need increasingly empowerment and support in this process, a new and challenging responsibility for nurse management in today's NHS climate. Increasingly, tiered management structures have been replaced with decentralised patient-centred and clinically relevant management. Within this arena three important new concepts for nursing have emerged.

Accountability

The Nurses Code of Professional Conduct[2] requires that each Registered Nurse, Midwife and Health Visitor is accountable for his or her own practice. A fundamental change within nursing in recent years has been the move away from the subservient role of the nurse. No longer do they unquestioningly deliver all medical prescriptions for care, whilst simultaneously being controlled by nursing management hierarchy in matters of personal conduct and discipline. The emergence of nursing as a profession implies a contract between the public and the nurse. In return for status and recognition, the public demands a standard of practice to which all nurses will aspire. The Code of Conduct sets out the principles by which nurses work in the exercise of professional accountability.

With the development of present-day management structures the 'to whom is the nurse accountable' question is often raised. Nurses may find themselves facing in different directions, with professional accountability through the Nurse Manager to the Director of Nursing and managerial accountability within a clinical specialty or directorate to a non-Nurse Manager. Primarily, of course, the nurse is accountable to patients and in the exercise of this function nurses have sometimes found themselves in conflict with management. For example, where insufficient resources are available or the nursing environment is inadequate to maintain quality patient care. This conflict of interests or accountability may increasingly be a problem for nurses practicing their profession within the resource limited NHS. Whilst present-day nurse education programmes prepare nurses to act as increasingly independent practitioners, those who undertook their training in a different era, and most especially those 'returning' to nursing, find the issue of accountability a difficult and a threatening one to handle.

Competence

Arising from the concept of accountability, 'competence to practice' also needs consideration. At the time of registration the nurse will have satisfied holders of the United Kingdom Central Council (UKCC) Professional Register that they have acquired the competencies of a first (RGN) or second (EN) level nurse. These competencies are set out in statute.[3] In the post-registration period the responsibility for continued development and maintenance of competence is the responsibility of the individual nurse together with the Nurse Manager in whose unit the new nurse practices nursing skills and uses his/her knowledge. For nurses returning to practice following a break in their professional career there is the need for an assessment of their competence. Currently there is no national standard for this and it is left very much to the local Nurse Manager. The introduction of periodic re-registration for nurses' demands that they provide evidence of ongoing education, training in clinical skills and therefore proof of competence to practice. Nurses are now gathering information for their personal professional profiles for this purpose.

Continuing responsibility

The concepts of 'continuing responsibility' and 'continuing overall responsibility' were first introduced with the nurses' clinical grading exercise 1988. The exercise was intended to assess the value of each nursing post using a set of predetermined criteria, assign a grade for the post and, in so doing, provide a clinical career structure for nurses and an appropriate level of remuneration. The exercise proved to be extremely traumatic, professionally damaging and altogether totally demoralising for many nurses who, in the expectation of finally receiving recognition and financial reward for their skills, knowledge and experience instead found themselves in conflict with their peers and their Managers. The National Appeals mechanism continues to handle, four years on, hundreds of claims from dissatisfied nurses.

The exercise has, however, sought to clarify amongst others the responsibilities of the Ward Sisters and Nurse Managers. Such posts carry a responsibility for the ward or clinical unit throughout 24 hours. Ward Sisters are responsible, with their team, for assuring the assessment of patients' needs, the preparation of individualised care plans and the implementation and evaluation of those plans. Additionally they are responsible for setting and maintaining standards of care within their area, ensuring that staff are appropriately deployed on all shifts and that policies and procedures are in place to ensure the ward operates to a high standard, whether or not they are on duty. Similarly 'continuing overall responsibility' lies with the Senior Nurse or Nurse Manager with several wards in a clinical unit.

The grading criterion was quite explicit in that there could only be one such post within a ward or unit; the responsibilities could not be shared, although job sharing was not precluded. Whilst a majority of Ward Sisters and Senior Nurses have had no difficulty with this shouldering of responsibility and the need to be accountable for it, others have struggled and some fail quite miserably.

The role

It is against this backdrop that today's Director of Nursing is working to give guidance, direction and leadership to the nursing profession and advice and support to the medical profession and those individuals who make up the National Health Service Management Team. The role of Director of Nursing may vary and is, of course, subject to a number of external influences. The job description, the expectations of other officers, the person to whom you are accountable and, not least, the demands and expectations of those nurses and Nurse Managers with whom the Nursing Director works. Additionally it will reflect the individual's own interpretation of the role — their ideologies — together with an understanding of what is required in their particular organisation.

Four different models appear to emerge:

Professional leader

The Director of Nursing is seen as a role model for other nurses working within an organisation. They possess strong leadership qualities and have a clearly defined sense of the purpose and strategy for nursing. They command respect and loyalty and are able to motivate large numbers of nursing staff as they strive towards the provision of high quality care. They establish effective working relationships with nurses of all grades and disciplines and additionally with colleagues in other professions. Their dilemma is one of how to maintain clinical and professional credibility. How to continue to identify with nurses at patient care level when the time for spending in this activity is so limited. Furthermore, how to maintain their knowledge of the many disciplines of nursing and its specialisms in such a way that they are able to be effective in their role. In this respect the Medical Director who continues in a clinical role as a Consultant has a clear advantage.

Adviser/representative

The Nursing Director will hold a statutory nursing qualification and may or may not be involved with issues of nursing practice and responsibility for standards of care. They will, however, be the representative of nurses, ensuring that their voice is heard in discussions and decision-making at management level. They will be severely hampered in this if they are not in touch with current issues in nursing development and practice. The Nursing Director will provide advice to medical colleagues and those of other disciplines on issues which arise. They will communicate with Nurse Managers within the organisation regarding professional matters and may act as Chair for the Nursing, Midwifery and Advisory Committee. In this model the Director may, and frequently does, hold other management responsibilities.

Manager

The Nursing Director will have defined management responsibility. They will be the budget holder and functional manager for all nurses, providing the nursing resource as defined by a particular clinical specialty or department. They are responsible for recruitment and selection, education and professional development, management and disciplinary matters. Alternatively, the responsibilities of the Director are combined with those of operational management, clinical service groups and directorates being managerially accountable to the Director of Operations, who happens to be a nurse who can double as the Director of Nursing. The dilemma here must inevitably be a conflict of interest. Frequently responsibility for ensuring quality within the organisation falls to this manager. This seems eminently sensible. The professional leadership role and the quality role are comfortable bedfellows. So much of patient care is influenced by the attitudes, the behaviour and the practice of nurses that there is much to be said in favour of this particular combination of roles.

Director/executive

Within the NHS reforms the role of Executive Director of Nursing has been identified. In 1990 NHS Trusts were established by the Secretary of State.[4] The Executive Director of Nursing is a member of this Corporate Body, however, little guidance is contained within the Act regarding the responsibilities and liabilities of Directors. The general function of the Trust is set out:[5]

> ... an NHS Trust shall carry out effectively, efficiently and economically the functions for the time being conferred on it by an order under Section 5(1) of this Act.

It is in the light of this general function that the Director of Nursing must work out particular responsibilities and clarify and demonstrate a clear understanding of obligations within the Trust.

It is of paramount importance that the delivery of nursing care is in the hands of competent practitioners and nursing managers who have a clearly defined framework in which to practice, where protocols and policies and procedures are clearly set out. The Director of Nursing will influence the decision-making processes to ensure that adequate resources, both financial and human, are available to deliver comprehensive and quality nursing care. In this respect the Director of Nursing may discover the same conflict of interests as more junior nursing colleagues. Conflict may arise between the duties the Director has as a member of the Trust Board, particularly in financial management and their professional duties as a nurse. Such conflicts are often perceived rather than real. In situations where inadequate resources appear to be available to meet a particular need, looking around there is frequently evidence of resources being squandered. For example, in circumstances where a sufficient number of nurses are not available on a particular shift, a study of the nurses' off duties will indicate that too many have been allowed time off on a particular day, that there is an inexplicable and unnecessary overlap of shifts or that sickness absence is both unmonitored and unmanaged. Whilst unpalatable decisions involving change in working practice may be required, the Director is failing in both professional accountability for patient care and accountability as a Director if he/she shrinks from these.

Whilst the role may be predominantly one or other of these models in most cases elements of all will be apparent.

The relationships

Board or management team

There are several crucial working relationships for the Director within a Directly Managed Unit or Trust. Paramount within these will be the working relationship with the Chief Executive or General Manager and with the Medical Director. This triumvirate, working harmoniously and effectively, will set a pattern for equally effective management throughout the organisation. Nurses need not be fearful of the clinical directorate model of management, rather they should feel liberated by it. Most nurses, once registered, identify within a particular clinical specialty. Management responsibility and accountability at this level allows for a contribution to problem solving, decision making and the planning of developments for nurses who are concerned on a daily basis with the delivery of patient care. It gives the opportunity together with medical colleagues for the involvement of professionals in management so forcefully argued by Griffiths.[6] A clear definition of the working relationship between Chief Executive or professional manager, the Medical Director and the Director of Nursing, where professionals are taking the lead in planning, developing and delivering healthcare services and management is facilitating and enabling. This process is a recipe for success.

Both the Director of Nursing and the Medical Director are primarily professionals, although they may lack the financial, statistical, technical and managerial finesse needed at Board or Unit Management level. Equally, they may not. Certainly through the combined expertise of Information Manager, Director of Finance and the General Manager, together with the skills, knowledge and experience of medicine and nursing, they will form a highly effective team. In the relationship with Executive and Non-Executive Directors or members, the Director of Nursing will seek to advise and influence decision making in favour of high quality care. The Nursing Director will need a wide-ranging knowledge and ideally experience of each nursing discipline and its role. They will need to speak with authority and an awareness of current developments, influences and trends on matters of nursing and patient care.

Marginally nurses have had greater opportunity than their medical colleagues to develop management skills. Doctors, though, are quick and eager to learn. Many of the skills developed through the management of care and treatment will prepare them for this new role and responsibility. The wise and effective Director of Nursing will form a working relationship with the Medical Director, which is both supportive and collaborative. There will be circumstances in which you will prompt action by the Medical Director and circumstances where action by either Medical Director or Clinical Director will be confronted. Above all, though, these Directors will be professional allies who share a determination that their organisation should provide all patients with high quality care and its staff, of all disciplines, the satisfaction of being valued and enabled to do a worthwhile job well. The nurse is, of course, no longer the hand-maiden of the doctor, at Board or indeed at any other level, although this may yet need to be grasped by some doctors and not a few nurses. The two professions complement each other and together they are a formidable force in leading the Health Service forward into the twenty-first century.

Clinical directorates

Having established the management model for the organisation with the Chief Executive and Medical Director, the relationship with Clinical Directors is an advisory one. The Nursing Director will reinforce the management responsibility of the directorate for those staff and resources within their remit and will act in such a way as to reinforce the lines of managerial and professional accountability for nurses within the organisation. In circumstances where the Director of Nursing has managerial or operational responsibility that can override the managerial accountability of the Directorate or clinical service group tensions may well arise. The Director of Nursing is a resource to Clinical Directors and may be used to give advice, or act as a sounding board for discussion regarding disciplinary matters concerned with nurses or indeed other staff. Clinical Directors may seek guidance regarding standards of practice or the appropriateness of allocated resources looking for backing as they progress their argument in support of growth and development. Take care in remembering that the Director of Nursing is not the manager of nursing, although they are their professional leader.

Nurse managers

Whilst the role of Matron has experienced metamorphosis over two centuries, middle grade nurse management has suffered an even more severe identity crisis. Even now there is a variety of models. In a recent report the Audit Commission[7] comments:

> Nursing management varies immensely in structure, in style and in operational matters from one hospital to another. No single solution or pace of change is right for all hospitals but whichever path is chosen will need to promote the ideal of clinically relevant management led by patient needs. This requires strong leadership, good communication and a well educated nursing force.

Traditionally decisions about the delivery of patient care, the management of the nursing workforce and the deployment of resources have been fed up the nursing hierarchy. In the here and now the buck stops within the directorate or clinical service group with a minority of issues requiring the intervention of the Director of Nursing. Nurses at middle management level will need appropriate preparation for this new role. They will be selected for their ability to make sound professional decisions, take appropriate action and seek guidance and advice in the face of complex situations. They will be confident about good modern day clinical practice and prepared to confront traditional methods and working practices, able to manage change. In return they will need support, encouragement and a clear strategy within which to work; and lots of well dones!

Whilst Nurse Managers will look to the Director of Nursing for leadership, advice and counsel, and whilst they will work closely together on all professional issues, they too are managed within the directorate either by the Clinical Director or, alternatively, through the Business or General Manager. This again is a potential area of conflict. The Director of Nursing will do well to define role relationships clearly with all colleagues and not stray from them.

External organisations

The Director of Nursing will provide a focal point for liaison with a number of outside agencies.

Colleges of health

Concerned with the preparation for registration and post-registration education for nurses. Directors of Education will need a focus for planning and discussion within the Trust or directly managed unit (DMU). Contracts will exist between the two for the provision of educational courses and reciprocally for clinical placement experience for nurses in training.

General practitioners/FHSAs

The debate continues still over the purchasing and providing of primary nursing services. Whatever the outcome the two agencies, Health Authority and Family Health Service Authority (FHSA) will look to Directors of Nursing to influence the provision of care by district nurses, health visitors, school nurses and family planning nurses. With the shift in emphasis to care at home the Director of Nursing will need to understand fully the cost and consequence of such a change if they are to be effective in their role.

Health authorities

As the quality of care is increasingly defined within contracts with purchasers, the Director of Nursing will provide a focus for debate. Provider units will be expected to identify and develop indicators and outcome measures relating to quality care. A current example where dialogue and collaborative working and monitoring is required is the Patient's Charter[8] demand for 'each patient to know the name of the nurse responsible for their care and how to contact them'.

The responsibilities

A prime responsibility within either a directly managed unit or a Trust is ensuring that attention is focused on strategic issues, policies and priorities. Ensuring that quality services for patients and people are always a major factor in decision making and service delivery. Increasingly within the contracts debate, the competing pressures of cost and volume of activity will attempt to force a quality compromise. The Medical and Clinical Directors and the Director of Nursing and Nurse Managers must be the guardians of quality defining the framework for care and defending standards in the face of increasing pressure to undertake more work with the same resources or, even more critically, with fewer resources. Nurses and doctors are at the interface with patients. Nurses are the only professional group who remain at the patient's bedside 24 hours a day. So placed they are able to identify unacceptable standards and potential deterioration in care quality. The Director of Nursing must ensure that they are enabled to express their concern and that information systems are in place to provide good quality and supportive data for their arguments. Together with Nurse Managers, specialist nurses and Ward Sisters, the Director of Nursing will need to define 'a strategy for nursing', which gives consideration to the following factors:

Philosophy

The strategy of nursing provides a framework in which nurses can practice, setting out clearly the goals and objectives and where change is required, thereby enhancing the possibilities of success. The strategy will reflect the purpose and values of nurses and nursing, defining patient-related and nurse-related values:[9]

> The purpose of nursing, midwifery and health visiting is to assist each person to obtain, retain or regain their optimal state of health and independence within the limitations of any physical, psychological or social constraint. This includes preparing individuals to cope with and assisting them to deal with major life events such as childbirth and bereavement and helping them to live with the effects of illness, disease or disability. When recovery of health is not a realistic goal the purpose of nursing is to ensure that the patient receives appropriate compensatory care and support aimed at achieving the best attainable quality of life or at the end of life to support and comfort the dying in order that death may be peaceful and dignified.

Clinical practice

Throughout the interface of the medical and nursing profession as, doctors have developed their skills, knowledge and ever-increasing complex technical abilities, nurses have been extending their role to take up elements of care previously undertaken by medics. The trend continues. Nurses are increasingly encouraged to expand their clinical practice in a variety of ways and become evermore skilled and specialist in their own right. The current development of the role of nurse practitioner is a further extension of this process and will allow further frontiers to be crossed by nurses in the delivery of care. The issues of professional accountability and competence already discussed earlier in this chapter set the framework for this development. The Director of Nursing will ensure that, where necessary and appropriate, protocols and operational policies associated with the development are drawn up. The Director will need to ensure that the management team or Trust Board, who has vicarious liability for all its employees engaged in the delivery of care is aware of this development and its consequences and is confident that neither patient, nor nurse or indeed management is being exposed to unnecessary risk. In the increasing debate regarding their eduction in junior doctors' hours the emergence of nurse practitioners is inevitable.

Care plans and models

The arrival of the nursing process from the USA in the early 1970s whereby a patient's individual care needs were assessed and a nursing diagnosis and treatment required were drawn up was heralded with concern and dismay amongst many, not least the medical profession. However, the subsequent development of care plans is generally welcomed, although perhaps poorly understood. The patient's individual needs are assessed by a Registered Nurse and a plan of care is then prepared, in which process the patient participates. Where appropriate the patient is helped to make choices in relation to the plan, which is then implemented. Evaluation should be a regular feature but it is in this aspect that, generally speaking, nurses have failed to exploit the value of the plan to the full. Evaluation skills related to the effectiveness of the care plan are poorly developed, consequently the care plan is often misleading in judging the quality and extent of care received by a patient. With the increasing commitment to the sharing of care between doctors and nurses questions are raised regarding the need for two separate assessments of the patient's condition, particularly following elective admission where diagnosis is already made and treatment already planned. The two professions need to develop a collaborative assessment and care planning process, so avoiding unnecessary workload for themselves and the trauma of two detailed sets of questions on similar themes for patients.

In the move from task-orientated to total-patient or 'holistic' care and alongside a need for nurses to define the unique role of the nurse, the concept of models of nursing has developed. Such models allow a systematic and organised review and assessment of patient need as a preface to preparing the care plan and the practice of nursing. Roper *et al.*[10] model uses 'the activities of daily living' as a framework for the nursing assessment. Twelve activities of living are described, they include, among others, breathing, communicating, eating and drinking, sleeping and dying.

Primary/team/task nursing

The tradition of the patient's bath book, the routine TPR rounds and the wound dressings all undertaken by the Ward Sister are thankfully now a thing of the past. Nurses have recognised the value of holistic care for their patients and have moved away from task-oriented care to team or primary nursing in the interests of continuity and quality of care. Within the primary nursing context the primary nurse is responsible for all aspects of patient care throughout the patient's stay in hospital. Team nursing works on the principle that a team of nurses share this responsibility, so ensuring continuity of care. Whilst patients like their care delivered in this way there is no evidence to support the view that this enhances the effectiveness of outcomes of nursing care. In the main, nursing opinion supports the view that improved communication, increased continuity of care and better informed decisions will all contribute to a more satisfactory outcome. The medical profession has, however, struggled with the concept of the primary nurse. Many continue to expect the Ward Sister to take the lead nurse role in the Consultant ward round and to be better informed than nurses within the team of patients' needs, problems and the care delivered. Why should not both the nurse in charge and the primary nurse be available with the patient during the ward round, so allowing the primary nurse to demonstrate knowledge of the needs and care of patients in their charge? With the Ward Sister exercising the role of facilitator, educationalist and the one with continuing responsibility for all patients within the ward.

Professional education

The strategy for nursing will define the need for ongoing professional education and training for nurses. Ensuring that skills and knowledge of their chosen field of work continue to be extended and enhanced. The UK Central Council has set out a framework for post-registration education and practice (PREPP). Ongoing registration for practice will be dependent on the nurse providing evidence every three years of further professional education.

Research

A further key objective within the strategy will be that of nursing-based research; this work supports the practice of nursing. Nurse management will ensure that the climate for research is right by encouraging nurses to question, monitor and review, incorporating subsequent findings into current nursing practice. This need not be a complex process but rather the development of an attitude that is constantly seeking ways of improving the quality of care delivered to patients.

Quality

Within the strategy the quality of care will be a central issue. Standards of care will need to be identified and a system of nursing audit set up so that standards are continually monitored and reviewed. Currently a number of nursing audit systems are available, i.e. MONITOR.[11] Increasingly nurses are developing tools for monitoring standards of care at local and 'grass-roots' level. This is to be encouraged. Such systems examine the framework of care, patient records and question the patient and their family regarding their experience of nursing care. Additionally, educational audits are frequently conducted within clinical areas. These assess the suitability of the learning environment for nurses undertaking pre- and post-registration education and training. Increasingly nursing audit will develop in conjunction with medical and paramedical audit, evaluating intervention and outcome where care is provided jointly. Simultaneously nurses are developing skills in the design of locally appropriate patient satisfaction questionnaires. These provide valuable feedback for the Director of Nursing and Nurse Managers regarding quality of care.

Manpower

The Director of Nursing will be expected by nursing and non-nursing colleagues to identify an appropriate establishment or number of nurses required to undertake the work within clinical areas. Additionally, judgements will have to be made regarding the skill mix. For example, how many first and second level Registered Nurses are required? What proportion of support workers should the team include? Do we include staff with clerical, domestic and housekeeping skills, so relieving nurses of the non-nursing duties that so often keep them from the patient's bedside? As the pressures of increased efficiency and maintaining low costs are felt, the nursing budget will be a target that will need an informed and confident defence. The provision of nursing services represents some 35 per cent of the total expenditure of the National Health Service. Therefore, in any assessment of the relative efficiency of health care, the proper assessment of the use of nursing resources is absolutely essential. The Director will be aided in the defence if a defined establishment, adequate to ensure a safe and acceptable quality of care, has been defined in a systematic way using proven methods and valid information. A number of methods are available to assist the Director in defining the establishment. None are foolproof and most can be questioned and challenged. None the less some markers will need to be set down:

- Nurse to patient ratios.
- Occupancy by specialty.

72 Nursing management issues

- Workload and dependancy analysis.
- Outcome measures.

Nurse/patient ratios

This rather crude estimate of nurse staffing requirements is perhaps the most traditional. The ratios in Table 4.1 were produced in 1984. The table also indicates the number of nurses required per bed in a particular clinical specialty.

Table 4.1 Nurse/patient ratios

	Nurses	Beds
General medicine and surgery	1	1:9
Paediatrics	1	1:2
Day care beds	1	2:6
Elderly care	1	1:3

These numbers are calculated giving whole time equivalent nurses required.

This figure is then increased by adding:

- A percentage for small wards, i.e. less than 30 beds.
- An additional 0.33 whole time equivalent for each learner nurse.
- A percentage allowance for approved leave.
- A percentage allowance for sickness.

The final figure represents the estimated establishment for a particular ward. This method takes no account of patient dependency, nursing need, occupancy and throughput. Additionally, the ratios of nurses to beds can be challenged. It must therefore be further informed by other methods.

Occupancy by specialty

Guidelines are available[12] that set out by specialty the number of nurses required. The figures are then adjusted according to the size of the ward and occupancy levels. For example:

General surgery

20 beds. 81–90 per cent occupancy. 18.80 WTE, including five learners. These estimates include a percentage for annual and sick leave, and for training. Additionally, advice is given regarding the allocation of learner nurses and night duty staffing.

Workload/dependency.

A number of methods using information on patient dependency have been developed. Where these systems are computerised the Nurse Manager will have operational information on a daily basis. This can be used to ensure adequate nurses are available on a shift and to aid the deployment and redeployment between clinical areas, as well as giving an indication of the establishment required. The South West Region is currently developing a patient-based system already operational in most of its District General Hospitals. Patients are assessed on a daily basis allowing a care group identification. The system is already primed with the time required for a range of nursing interventions and the skill level of the nurse. This information is based on the professional judgement of nurses involved in setting up the system within a clinical unit. Hours of nursing time available are fed into the system from duty rotas, the outcome being an estimate of the nursing hours, trained and untrained required. Overstaffing and deficits by day and night are indicated. This system allows a more accurate assessment of the establishment required. The system can be manipulated by an over assessment of patient need. For this reason regular audit is required, persistent offenders are easily identified!

Outcome measures

An estimate of nursing numbers required per available bed, further informed by clinical specialty, occupancy and throughput is valuable. If a system for determining individual patient need or workload is in being, better still. Increasingly, however, the outcomes of nursing intervention will have to be considered. More especially, what is the outcome of an inadequate number of nurses available to provide safe and acceptable care? Is it an increase in the incidence of pressure sores? Costly to say the very least. Is it a lack of time for the proper exchange of information both pre- and post-operatively between nurses and patients? The outcome of which may be increased anxiety, increased pain, slower recovery and a higher incidence of complications and readmission? These are unchartered waters. Nursing Audit is in its infancy, considerable effort is required in gathering meaningful information in support of adequate nursing establishments. This course must, however, be set and nurses prepared to take the helm if the unique contribution to patient care which is nursing is not to be compromised.

Developments in nursing

The Director of Nursing will strive to keep abreast of developments related to nursing and patient care, so ensuring effective professional leadership and prompting for nurses within the organisation. Additionally, the Director will advise colleagues of changes and their implications so that appropriate planning and preparation takes place maximising the benefits of any development.

Major developments currently being implemented include:

- Project 2000.
- Computers in nursing.

Project 2000

Major changes in the health care needs of the population have heralded the need for change in the way student nurses are prepared for nursing. A greater proportion of elderly people than ever before is creating an increasing demand for different health services. At the same time fewer young people are available for work, and there is a shift of health care provision into the community and an increase in health promotion and illness prevention activities. In the face of such change, Project 2000, a new programme of education for nurses, has been introduced. It is the profession's response to the challenge of providing nursing care in the twenty-first century.

The course lasts three years and, following a common core foundation programme, prepares nurses in one of four branches:

- General nursing of adults.
- Mental health nursing.
- Mental handicap nursing.
- Childrens nursing.

> The branch programmes prepare the student to apply the broad principles and skills in the common foundation programme to assess, plan, implement and evaluate the nursing needs of individuals and groups in health and in sickness, within their chosen area of practice. (Department of Health[13])

The responsibility of the Director of Nursing in relation to the introduction of Project 2000 is threefold:

(1) Recognising, valuing and enhancing the skills and experience of nurses trained under the traditional system in order to maintain the quality of patient care.
(2) The preparation of clinical areas and community care settings for the placement of Project 2000 students requiring experience of care delivery. If the new educational needs of these students are to be met much

change will be required. Existing Registered Nurses and colleagues in other disciplines will need to be informed of the new programme and its implications. 'Clinical' coaches whose skills in the clinicial areas will be required for coaching, mentorship and assessment of the 'new' students must be identified. Suitable clinical areas in which 'the rostered' contribution will be made, must be agreed between education and service managers.
(3) Replacement of the service contributions made by students on the traditional programme.

A fundamental change with the introduction of Project 2000 is the 'student status' of its participants. They are 'supernumary', receiving a bursary rather than a salary.

Whilst this new breed of student make a service contribution of 1000 hours towards the end of their course this represents a considerable reduction in manpower. Currently learners make a 60 per cent service contribution. The Director will be concerned with the replacement of this service contribution within strictly defined financial limits worked out by the Department of Health. It is in this scenario that the support worker or health care assistant is emerging. They will play a major part in replacing the service contribution currently made by learner nurses. The health care assistant works under the direction and supervision of a Registered Nurse. Most will have achieved National Vocational Qualification level 2 'Assisting Clients in Care'. With experience and training they will carry out many of the tasks associated with the delivery of care.

Computers in nursing

The role of computers in nursing has already been discussed in relation to the determination of an appropriate nursing establishment. Using patient-based information following individual assessment a detailed workload analysis is available. This information can also be used to identify tasks carried out in the delivery of care for which Health Care Assistants rather than Registered Nurses may be prepared in the future. At the same time, the computer can consider those patients who require exclusively the skills knowledge and experience of a Registered Nurse. Additionally, and increasingly, such computers are linked to Resource Management systems allowing the cost of delivering nursing care to be more accurately calculated in relation to the cost per case information required for contracting purposes. A comprehensive system will produce care plans, so reducing the amount of time spent by nurses in preparing these. It should also allow for the development of nursing audit and the gathering of nursing outcome data.

In summary, the Director of Nursing, who is to effectively lead, defend and deploy nursing resources, will require an information system that:

- Uses patient-centred information.
- Assesses workload.
- Determines establishment requirements.
- Is finance linked.
- Produces individualised care plans.
- Facilitates nursing audit.

Dilemmas

Throughout this chapter we have discussed in detail the role and responsibilities of the Director of Nursing, who treads a complex and often lonely path. There will be dilemmas.

Resource allocation

With the devolvement of budget responsibility to clinical service or directorate level the Nursing Director will need to ensure a fair allocation between acute and community-based services. So often the demands of acute services have been paramount, depriving the community of much needed resources. Increasingly care is moving away from large acute units and resources must follow. The Director has an equal responsibility for the safety and acceptable quality of care for patients in hospital and at home; few colleagues share this, they are concerned with one or the other. Similarly systems must allow for the redeployment of nurses across directorate boundaries when the need arises, for the sharing of equipment and other resources, the attitude of the Director can influence this factor greatly.

Consistency and standards

A similar overview will be required in relation to the standard of care delivery. The Director will have a role in overseeing ward environmental audit, educational audit and in prompting and developing nursing audit. They should also play a part in diverting resources where this is required in the maintenance of standards.

Control

The Director of Nursing, unlike the Matron, is no longer in control. The responsibility for the nursing budget and the management of nurses has been devolved to directorate level. They have the authority to act, the decisions now are theirs. The Director provides the framework in which all nurses practice and should ensure that the nurses are familiar with policy and principles for practice, that they have the resources they need and that they are aware of the standards required. The Director enables and facilitates an increasingly responsible and accountable nursing workforce in the execution of duty.

References

1. Dingwall R, Rafferty AM, Webster C. *Introduction to the social history of nursing*. London: Routledge, 1988.
2. United Kingdom Central Council. *Code of professional conduct*. London: 1984.
3. Nurses Midwives and Health Visitors Rules Approved Order, No. 873. London: HMSO, 1983.
4. National Health Service and Community Care Act, 1990. London: Department of Health, 1990.
5. National Health Service and Community Care Act, Paragraph 6, Schedule 2, 1990. London: HMSO, 1990.
6. Griffiths R. *NHS management enquiry* (Griffiths Report). London: HMSO, 1983.
7. Audit Commission. *The virtue of patients. Report of the Audit Commission*. London: HMSO, 1991.
8. Department of Health. *The patient's charter*. London: HMSO, 1991.
9. Department of Health. *The strategy for nursing midwifery and health visiting Scotland*. Edinburgh: HMSO.
10. Roper N, Wogan WW, Tierwey AJ. *The elements of nursing*. Edinburgh: Churchill Livingstone, 1990.
11. Goldstone LA, Ball JA, Collier MM. *MONITOR systems of nursing audit*. Newcastle-upon-Tyne: Newcastle-upon-Tyne Polytechnic Products Ltd, 1989.
12. Birmingham: West Midlands Regional Health Authority. *Nursing midwifery and health visiting staffing guidelines*. 1989.
13. Department of Health. *Health Circular 8240257 12/90*.

5

Managing audit
Tony White

Introduction	78
Historical background	79
Principles of audit	80
The audit cycle	80
Scope of audit	81
Essentials of audit	82
Organisation of audit	82
Involving others in audit	83
Confidentiality of audit	84
Definitions	85
Quality assurance	88
Health service indicators	88
Management and audit	89
Other forms of audit	90
Total quality management	92
References	94

Audi partem alteram.
(St Augustine, *De Duabus Animabus*, XIV. ii)

Introduction

Audit derives from the Latin, meaning 'hear'. According to the Oxford Dictionary, the term 'Audit' means an official examination of accounts, but The Royal College of Surgeons defines it as the systematic appraisal of the implementation and outcome of any process in the context of prescribed targets and standards.[1]

Audit is now a word much bandied about by politicians, managers and doctors. It usually has a prefix attached and there seems to be some confusion about what they all mean. Many managers with whom I have spoken admit privately to ignorance of audit, especially medical and clinical audit, and even candidates at consultant interviews, when asked about audit, appear to have confused notions.

There are now many books, papers, conferences and courses for doctors, managers and other professionals on how to conduct audit and it is not my intention to duplicate those many excellent sources of information. I would, however, like to clarify a number of definitions that appear to be used indiscriminately and, more importantly, to highlight some of the ways in which the process of audit can be used more effectively.

Medical audit should primarily be an educational activity and has to be professionally led because it is a peer review of medical practice, ensuring that the quality of medical work meets acceptable standards. It is therefore necessary to have and to understand current medical practice. This is recognised in the White Paper,[2] which states: 'medical audit is essentially a professional matter'. The purpose is to improve the standards of patient care. Audit, unlike research, does not necessarily extend the knowledge base of medicine but, by critically analysing medical practices, aims to improve the quality of care.

Historical background

There has always been a tradition in medicine of considering the process and outcomes of clinical practice to improve patient care. To some extent the current ideas about medical audit are steeped in that traditional view. It is the use of the term audit that is relatively new and somewhat threatening to the medical profession the term has now become an accepted part of health service jargon. Yet the idea of medical audit is not new, reference to it can be found in the Charter of the Royal College of Physicians of 1518, which states that one of the College's functions is to uphold the standards of medicine 'both for their own honour and public benefit'. In 1952 the Department of Health and the Royal College of Obstetricians and Gynaecologists[3] began the Confidential Enquiry into Maternal Deaths. The Royal College of General Practitioners was involved early on in the medical audit of general practice. Much of the work of the Birmingham Research Unit in the 1950s and 1960s was concerned with audit. In 1969 the National External Quality Assessment Scheme was started involving all the commonly used investigations in pathology. In 1980 the Royal College of Physicians conducted a survey of causes of death in medical wards of all patients under 50. In 1983 The Royal College of General Practitioners began its quality initiative,[4] introducing a new system for entry to its Fellowship based on performance based assessment in practice. Beginning in 1989 an enquiry by anaesthetists and surgeons into perioperative deaths in three regions (CEPOD)[5] was extended into a national enquiry.

The Royal Colleges and their faculties have always taken a keen interest in the principle of medical audit and increasingly seek evidence of functioning audit before accrediting posts for specialist training. Kenneth Clarke, when Minister of Health in 1989 hailed medical audit as 'one of the chief means of raising standards of care'. The Government, having recognised that quality of patient care could only be achieved with the cooperation of doctors, established the Clinical Standards Advisory Committee, taking advice from the Royal Colleges, the SMAC produced a report[6] published in October 1990. This traced the historical development of medical audit and made recommendations on putting it into practice. It concerned itself solely with medical care as provided by the medical profession, which it described as being commonly referred to as 'medical audit'. The Government stated that all doctors were expected to be participating in medical audit by 1991 and detailed its expectations of regional and district health authorities as well as the clinicians.

Principles of audit

- An acceptance by medical staff of corporate responsibility for audit of the quality of care.
- A forum to receive information, promote discussion and make recommendations.
- A mechanism for deciding where responsibility lies between management and medical staff where this is uncertain.
- Recording of findings.
- System for acting on results.
- A mechanism to re-audit as commitment must be continuous; audit is a never ending process.

The audit cycle

The audit cycle is essentially concerned with changing practice to improve patient care. The importance of going round the audit loop has been stressed by many authors but there can be reasons for simply collecting data with the purpose of answering a question that concerns you, and only later going on to set standards and monitoring. The emphasis on the audit cycle is that by setting standards and monitoring the areas that require change may become obvious.

Highlighting those areas where a practice, as a group or as an individual, has not met the standard you had previously agreed, may give ample topics for discussion. The focus of implementing change occurs after the comparison of practice and standards has been made. The audit cycle itself does not tell you how to do this, or how to ensure that everyone is willing to become involved in the change and give it their full support. That issue is addressed in Chapter 7. However, at this stage it is important to realise that to reap the benefits of audit it is important to complete the full cycle and close the loop.

```
           Select Topic
          ↗            ↘
                 Observe    Define
                 practice   standards
                    ↓          ↓
Implement change  Define    Observe
                  standards practice
          ↖            ↙
              Compare ←
```

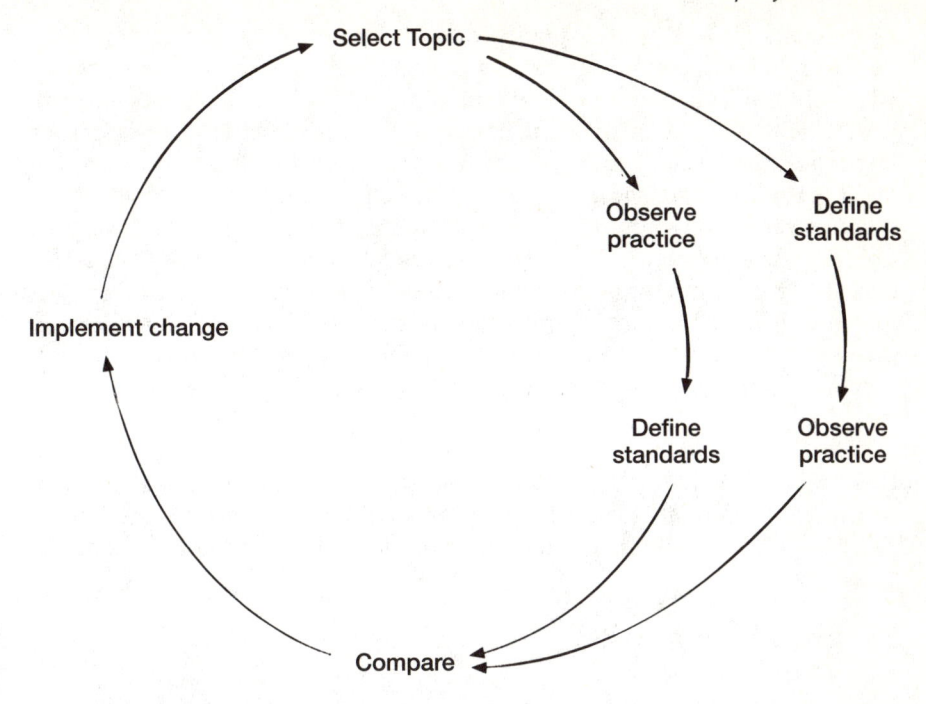

Scope of audit

There are many ways of considering the scope of audit but a useful classification is to consider audit under the following headings:

Structure

This would include such topics as appointments, admissions, waiting times, cancellation, non attendance, re-attendance rates and so on. The setting of targets to ensure equity of care, particularly where there are waiting lists.

Process

This would cover the use of resources such as out of hours operating or facilities for investigations. Clinical process could include the appropriateness and complications of clinical investigations or treatment. Administrative process would include effectiveness of communications and patient information, including medical records, referral letters or discharge letters.

Outcomes

Traditionally deaths and complications have long been considered by the profession. However, the quality of life after treatment, the degree of patient relief and comfort as well as the satisfaction of the referring doctor all need to be considered. One might consider positive outcomes, such as post-operative pain relief, or negative outcomes, such as post-operative infections and complications, pressures sores, etc. The *Report of a confidential enquiry into perioperative deaths* (CEPOD)[5] and Gruer et al.[7] have demonstrated the differing results between various units with varying expertise or workloads. The measurement of outcomes is seen as crucial for measuring quality and in particular achieving cost-effectiveness. For doctors this means that any move towards managed health care will mean providing the best and not just the cheapest care. In the US, medical outcomes studies have increasingly recognised the need to involve doctors.

It is probably most useful to audit those situations that are in one or more of the following categories: high volume, high cost, high risk or where there are wide variations in practice or perhaps some particular anxiety.

Essentials of audit

Many authors have summarised the essentials of audit but it might be useful to summarise them by asking a number of key questions:

- What should we be doing?
- What do we think we are doing?
- What are we doing?
- What are the results of what we are doing?
- How can we improve on that?
- Have we improved?

It is by finding the answers to these key questions that we can improve things.

Organisation of audit

It is normally one of the responsibilities of the Clinical Director to organise the audit within a directorate, unless it has been agreed that this has been delegated to another specified consultant. Speciality audit should be the responsibility of individual departments or divisions. At district level, audit would normally be medically led with a local Medical Audit Advisory Committee (MAAC) chaired by a senior clinician, with representatives of the major specialities including general practice, together with the Clinical Tutor, a Public Health Physician, a junior doctor and a representative of general management. This committee would be able to advise, encourage and monitor audit programmes, approving methods used and the frequency of audit for each clinical team. The District MAAC normally reports to a Regional Audit Committee and liaises with the Regional Specialist Sub-Committees. Liaison is particularly important because audit should serve the dual purpose of professional development and improving patient care, and to achieve these aims, audit has to be integrated within a wider framework. Audit may also serve other purposes, such as accountability and control of doctors and their activities. Defining audit in terms of peer review and primarily using it as an educational activity, removes some of the more threatening possibilities of outside control. The more audit is allowed to become external rather than internal the more threatening external judgements are likely to seem to the profession. The problem seems to be that government, managers and doctors all have differing agendas for audit. The important lesson is for doctors to embrace audit whole-heartedly before external authorities do it for us.

Involving others in audit

While it is usually easy to change one's own practice as a result of audit, improving clinical care often involves a number of people, colleagues, junior doctors, other professionals and managers. Encouraging others to change their behaviour often needs particular actions. It is also important to distinguish between auditing one's own work and that of others. In principle you should only ever audit your own work, although sometimes it is impossible to avoid looking at what colleagues do. Auditing others without their involvement leads to a number of problems, such as what to do with the data you collect. It is also imbued with the distinct possibility of suffering unexpected flaws of which you may be unaware. For example, some physicians were auditing the turn around time and service they were getting from a pathology laboratory because of delays in receiving investigation results. Thinking that the delays were due to poor service from the unit, the physicians simply arranged for the services to be supplied by another laboratory. The physicians were auditing something over which they had no control; the power to bring about real change lay with the pathologists. If the physicians had consulted and discussed the problem with the pathologists from the beginning, encouraging a joint audit and development of solutions, there would have been no need for their somewhat high-handed action. In this case, the physicians' 'solution' could potentially have a serious economic effect on a hospital's laboratory service.[28] It is vital to anticipate at the time you design your audit those who might be involved in future change and so encourage them to join you in designing it right from the beginning. Audit can also throw up surprises, which will often result in the need for changes that you have not anticipated. Managers, Local Audit Committees, Medical Audit Advisory Groups, Regional Audit Committees and Regional Specialist Advisory Committees all need to feel involved and will want to become a part of the process of any change. Managers should be encouraged to sit on the Local Audit Committee. Groups who do this have reported change being facilitated. Problems should be shared with management. Auditing audits that have indicated the need for management action, and the degree to which this management action has occurred, are a very useful excercise. A barrier to improvement may be the direct lack of involvement of other groups. Meetings with other specialty groups encourages audit to be seen as an educational activity and has many rewards in terms of improved care. So never forget the involvement of other professional groups.

Confidentiality of audit

Findings of individual cases need to be confidential, and issues relating to this need to be considered separately for patients, doctors and managers.

For patients, personal health information is confidential. It is necessary to anonymise patient data for audit meetings, and only summarised data and general conclusions should be passed to managers or Health Authorities.

For doctors, open discussion during peer review is essential and, to ensure this, it is vital that records are anonymised. Working proformas used for recording data need not be retained as they only duplicate information already in patient's records. Any serious professional problem identified can be dealt with within already established procedures — through the Departmental, Faculty or Unit Clinical Director, or the District Medical Director and ultimately the Regional Medical Director.

For managers, it is important to ensure that adequate medical audit procedures are in place and that adequate resources are available to support this. They will need reports that are anonymised. They will also need conclusions. In addition, recommendations, action plans and proposals for change where indicated and a note of when a review of the results of change should be made and the proposed method of review. The lessons learned might deserve being published more widely and some criteria for this have been published.[8]

The fundamental principal is that the quality of medical work can only be reviewed by a doctor's peers. It is a means of assessing and maintaining standards and the overall aim must be to ensure that patients receive the best care possible from the resources available.

Definitions

There are various terms over which there is some confusion, in particular medical audit and clinical audit, which are often used interchangeably, for instance Cookson in an article on computer technology and improving the use of hospital resources appears to make no distinction between medical and clinical audit,[9] so that it might be helpful to consider some definitions.

Medical audit is a systematic approach to peer review, where it is hoped that it can identify the possibilities of improvement in care and facilitate those improvements. It relates to the medical practice of doctors themselves. Clinical audit, however, covers all aspects of clinical care, including, for example, the work of nursing and paramedical staff. It is possible for clinical audit to overlap financial audit, utilisation review and management of resources, but it is primarily clinical and not managerial.[10]

Medical audit

The White Paper[11] defined medical audit as a systematic, critical analysis of the quality of medical care, including the procedures used for diagnosis and treatment, the use of resources and the resulting outcome for the patient. It continues that it is essentially a professional matter and is a means of ensuring, through peer review of medical practice, that the quality of medical work meets acceptable standards. It requires both specialised knowledge of current medical practice and access to adequate medical records. As the White Paper says, medicine is an inexact science, often lacking generally accepted measures of benefits, and therefore the audit process should not discourage doctors from taking on difficult clinical work.

Charles Shaw,[12] in a paper for the King's Fund Centre, defines medical audit as a systematic approach to peer review in order to identify opportunities for improvement and to provide a mechanism for bringing them about. He suggests that it complements and ultimately overlaps financial audit, utilisation review and resource management, but differs in that its purpose is primarily clinical rather than managerial. As Shaw says:

> Its focus is the process and results of medical care rather than resources. It is more systematic, quantified and formal than traditional clinical wards round, meetings and case presentations but shares with them the objectives of better patient care and post graduate education.

The Standing Medical Advisory Committee on The Quality of Medical Care[6] states that there are several definitions of medical audit but accepts that given in the NHS Review White Paper.[11] It states that the objective is to improve the quality of medical care doctors provide for patients, it is therefore an essential component of quality assurance of health care. It provides a systematic approach to peer review of medical care and its content is primarily clinical and educational, as opposed to managerial. Its focus is on the process and outcome of clinical care.

Clinical audit

The Standing Medical Advisory Committee[6] suggests that clinical audit be used to embrace the activities of all health care professionals who work directly with patients, such as nurses, doctors and paramedical staff. After all, the quality of medical care is not solely determined by doctors and nurses. Other health care professionals also have an essential role.

Surgical audit

The Royal College of Surgeons of England[13] has shown that it has already taken a lead in auditing surgical outcome, through its Quality Control Committee. For some time it has also been official College policy to make regular audit a condition for recognition of hospitals for surgical training.

Regional audit

A further development is described by Frater and Spiby,[14] who call the process whereby several districts within a region compare their performance 'global audit'. They describe the process of a group of clinicians pooling their data, discuss nationally agreed standards from the published results of consensus working parties, reported results of randomised controlled trials or other sources. Especially important for those smaller specialities where there may only be one consultant working single-handed in a district, global audit is more time consuming for any participant as more travelling and meeting time has be allocated in an already overcrowded schedule.

Financial audit

Shaw[12] suggests that medical audit complements and ultimately overlaps financial audit, utilisation review and resource management but differs in that its purpose is primarily managerial rather than clinical, its focus is the resources rather than process of and results of medical care. It reveals in conjunction with medical audit where deficiencies in care can be attributed to lack of resources.

Multispecialty audit

Also known as multidisciplinary audit, where specialties that actively involve other disciplines in clinical work or in a clinical team, e.g. surgeons and anaesthetists or psychiatry and geriatrics may also need to combine for some audit work. The enquiry by surgeons and anaesthetists into perioperative deaths is a good example.[5] The involvement of other professional and paramedical professions is a logical development from this but is then termed clinical audit. The Health Advisory Service set up in 1976 is an example of multidisciplinary audit. It carries out reviews of hospitals and community health services provided for the elderly and the mentally ill and makes recommendations for the improvement of care.

Audit of audit

The King's Fund Centre in its Guideline's for Medical Audit[12] suggested that very few of the hospitals in the UK currently (at the time of the report) fulfilled all the criteria set out. Even if agreed locally to be desirable, they might not be immediately achievable but that they would provide a basis for discussion and for audit of audit.

Auditing quality of service

The need for reviewing clinical work is crucial to any comprehensive programme for quality assurance. Bottomley[15] says that this was one of the five areas that GPs would have a vital contribution to play, as 'monitoring the quality of services'. She went on to explain that this would be reflecting not only their own assessments but also their patients' experiences.

This concerned not just the outcome of the process for the patient but his or her experiences while undergoing that process. For example, not keeping patients waiting in dreary outpatients rooms and greeting them on arrival in a friendly helpful way. Another important aspect of quality was ease of access to services.

Self-audit

The term 'self audit' may, be preferable to 'external review' because it emphasises the faults possibly inherent in data collection by non medical staff. There is ample evidence that data collected by clerical staff for HAA (hospital activity analysis) can be seriously inaccurate.[16-18] Interestingly they also indicate that events over the previous few years such as changes in the financial fortunes of the NHS have led to a closer inspection of all aspects of spending.

Criterion-based audit

Another name for occurrence screening or clinical outcomes programmes.

Occurrence screening as a method of audit is well described by Bennett and Walshe.[19]

Computerised audit

Since 1988 GPs have had access to data from Prescription Analyses and Cost (PACT), a system that allows analyses of all prescriptions issued in each quarter. All GPs automatically receive PACT level 1 data, a simple 4-page analysis. PACT level 3, which is much more detailed and gives the identity and price of every prescription item, is a powerful tool for auditing prescribing.[20] Up to two-thirds of GP prescriptions are issued as repeat prescriptions and computerised audit can help to monitor this.[21]

Clinical outcomes

Districts set themselves targets in terms of outcomes that are measurable in terms of improvements in the health of their populations.

Quality assurance

Medical audit is an essential component of quality assurance of health care and quality assurance is an essential part of the management process. It requires the definition of standards, the measurement of achievement and the setting up of mechanisms to improve performance. Quality assurance is a system by which provision or performance is measured against expectation with the declared intention of minimising deficiencies.[6] Clarke[22] stated that in his view 'medical audit is about quality assurance in clinical work'.

Quality assurance relates to accessibility, availability and equity of provision, the effectiveness in achieving this, and its efficiency and acceptability. There are, of course, difficulties in measuring effectiveness, neither can national, regional or district authorities control quality continuously, but peer review does help to 'assure quality if it is regular, objective, explicit and effective. That is medical audit.'[23]

Health service indicators

The Department of Health publishes health service indicators that include measures of activity, and some of outcome, e.g. stillbirth rates, neonatal mortality rates, low birth weight rates, post-neonatal mortality rates, perinatal mortality rates, immunisation rates and infectious disease notification rates.

There are also comparisons of death rates by region and district for certified causes of death from conditions considered potentially avoidable such as asthma, tuberculosis, hypertension, cerebrovascular disease and cancer of the cervix.

Management and audit

Many doctors have doubts, reservations and are suspicions about the role of managers and audit. It is important that those doubts and the uncertainties that accompany them are recognised openly. Audit is, after all, not a new idea or activity for the doctors, and the Government has placed the responsibility for medical audit unequivocally on the profession.[24] 'The quality of medical work can only be reviewed by a doctor's peers.'[29] Indeed the lead given by the Royal Colleges reinforces the professional ownership of audit. Nevertheless the development of medical audit in practice has so far appeared to be very largely dependent on the enthusiasm of individuals and there is little consistency of effort. Medical audit is an integral part of the NHS reforms and the SMAC views medical audit as 'part of normal clinical practice', stating that 'a failure to take part should be regarded very seriously'.[6]

The timetable of the NHS reforms has highlighted the need to plan and manage the development of medical audit at district level and hence there is an important role for managers. Managers, whether they are purchasers or providers, need information on the quality and outcomes of clinical services, as well as volume and costs, to manage resources to provide or purchase competitive services.

Managers also have to allocate sufficient resources for audit, whether this be allocating sessional time to assigned educational activities including audit, providing accurate and timely patient data, organising data collection (including regional and national data for comparisons) or retrieving patients' notes and records. They also need to provide secretarial support, stationery and computer facilities and allow clinicians adequate time and travelling expenses for Regional Audit and Specialist Subcommittee Meetings, etc.

Managers need to be involved in implementing the changes resulting from audit, whether that be in reorganisation of hospital practices, redistribution of staff or reallocation of funds. Many doctors still feel that managers will not be able to change things because there are no additional resources, but not every change costs money. Early treatment, avoidance of unnecessary or excessive use of drugs, investigations, etc. cost nothing. Avoidable delays in inpatient investigation, treatment and discharge; unnecessary outpatient reattendances and the cancellation of theatre lists, clinics and X-rays, etc. at short notice may cause unnecessary waste. But seeking new funds for extra manpower or equipment is an issue that has to be addressed by managers.

Managers need to integrate medical audit, quality assurance and resource management. And last, but by no means least important, they need to create and develop the culture of the organisation to share and understand the perceptions of medical audit, facilitate communications, manage the changes and be clear about various roles and relationships.

Managers in the role of providers of health care need clinicians to explain changes in practice, to evaluate new practices, to indicate dangers in existing practices and to prioritise demands on limited resources. Audit in itself provides an appropriate example, clinical workload may of necessity be cut to provide time for audit, without managers appreciating this unless it is explained to them. New techniques requiring investment in expertise and equipment may well have long-term benefits, such as more favourable outcomes and results, less patient time in hospital and less recurrent surgery, as well as being more acceptable to patients. Audit and the setting of standards will ensure that change produces benefit.

Managers as purchasers need doctors to explain quality matters, to advise on value (for money) issues and to allocate priorities relating to demand. Coordinated and integrated hospital and general practice audits can and will become incorporated into quality aspects of contracts. There are difficulties in achieving outcome measures so that much medical audit work to date has tended to concentrate on investigations and treatments. Increasingly, measurements of health gain will need to be specified in business plans and contracts, which will then be monitored, and it is this direction that purchasers will look to doctors to provide information. It is from audit that doctors will seek to provide this. This will require information systems to enable collaborative audit between hospital, general practice and community health services to take place.

Management audit must surely follow as a means of peer review to ensure that the quality of management meets acceptable standards. The Government's aim should be that all managers should participate. The work of management should be reviewed at regular and frequent intervals and although findings in individual cases should be confidential, the general results of management audit should be available locally and the lessons learned published more widely.

Other forms of audit

National audit

The role of the National Audit Office is unchanged. It is an Officer of Parliament, the Comptroller and Auditor General. It reports on value for money in the use of voted funds and certifies the aggregated accounts of the NHS, drawing upon the audits to be carried out by the Audit

Commission, but which have in the past been carried out by the Health Departments.[11]

The Royal College of Surgeons of England, commenting on National Audit of surgical practices, says this is necessary because districts are small units, each with a limited number of surgeons. The *National confidential enquiry into perioperative deaths* (NCEPOD) issues annual reports intended to improve standards of surgical practice. The reports include guidelines for specific clinical situations. The College also carries out quinquennial inspections by its Hospital Recognition Committee, which issues and regularly updates, recommendations about the training environment and facilities that each hospital provides. From January 1990 the Committee has initiated a routine scrutiny of hospital notes and audit records to ensure optimum standards of surgical care.[26]

Independent audit

The fundamental principal is that the quality of medical work can only be reviewed by a doctor's peers. However, the White Paper[11] says that management should be able to initiate an independent professional audit where necessary and that this could take the form of an external peer review or joint professional and managerial appraisal of a particular service. For example, where there is cause to question the quality or cost-effectiveness of a service. The possibility of independent doctors working jointly with the Audit Commission is also discussed.[11] The Government has proposed the transfer of responsibility for the statutory external audit in England and Wales to an independent body.

Government audit

The Audit Commission[3] is currently responsible for the audit of local authorities in England and Wales; these audits are carried out by a mixture of the Commission's own staff and private sector firms. The Commission has considerable experience and expertise in areas of work closely related to the Health Service. In particular, it is accustomed to working in multidisciplinary teams with professionals looking at the professional aspects of services. The Local Government and Housing Bill includes a provision that will enable the Audit Commission to undertake audit in the NHS under authority of the Secretary of State. This will develop the experience of the Audit Commission and enable its staff to start work with those officials in the Department of Health (DoH) and Welsh Office who are currently responsible for NHS audit. The Government has established the Audit Commission as the body responsible for the external audit of health authorities and other NHS bodies which are at present audited by the DoH and Welsh Office. It reports to the Secretary of State.

Total quality management

It is impossible to finish a chapter on audit without making some reference to total quality management (TQM). When things go wrong it often happens not within a single functional area, such as medicine, nursing, pharmacy, or administration, but often at the boundaries or interface between functions. Hospitals are divided into functions or departments, making it easy for people to blame each other (surgeons blame anaesthetists, and vice versa, doctors blame nurses, nurses blame technicians, technicians blame managers, managers blame doctors, one department blames another). It is difficult for any one group to see the process of work as a whole in the way the patient experiences it.

TQM theory directs attention to the processes of work rather than the workers. It postulates that most flaws come from processes, not people, it is the duty of managers to ensure that the process is designed and improved so as to permit the workers to do what they already want to do. The experience of process failure is frustrating and demoralising for people. These symptoms of frustration and low morale are seen widely in hospitals today. Health care is frustrated because it has not learnt how to get better. TQM, with its emphasis on continually improving the overall process, offers a mechanism for recovery. TQM is a continuous search for opportunities to improve. If the 'standard' complication rate is 2 per cent, people spend their time trying to get their rates below that. Those whose rates are 1 per cent have little incentive to make them even better. The result is mediocrity, or at least missed opportunities for improvement.

TQM requires a partnership of doctors, managers and other health professionals but to be effective it is necessary for doctors to understand and participate in management decisions, and managers to understand and contribute to the formulation of the goals of medical care. For this doctors will need to acquire new skills.

Interdependence and teamwork

Doctors may still bear ultimate responsibility for patient care but it is rarely true that patient care involves one doctor and one patient. Almost all of it requires collaboration with many health care professionals. Health care boundaries are cross-functional. As a result, internal dependencies develop. It is helpful for teams involved in a common process to meet as necessary to plan steps towards improvement.

The problem is how doctors might manage in such a culture of interdependences. Traditional medical roles and hierarchies of status are barriers for doctors. This is where the new skills are needed, to listen across the professional boundaries, to remain silent so that others of different status or background as well as colleagues can speak, and to realise that the idea of 'process' is a great equaliser.

Consultants need to show willingness to work effectively in teams, to share responsibility and to relinquish absolute professional autonomy in the service of shared purpose. Effective teams in health care will, in almost all situations, require active participation of doctors and frequently their leadership. Yet many doctors seem uncomfortable with real team activity.

Understanding process

In health care, 'process' usually refers to clinical care. In TQM it means the way in which work is done, and can refer to any type of work whether clinical, non-clinical or an interface between the two. Understanding process is an important step towards removing the fear of TQM. The fear of being made the scapegoat, of appearing ignorant or looking foolish. Fear can lead to defensiveness. When one blames processes and not people it should become safer to expose the facts and for everyone to get to work to improve the process. Seeing work in process terms can lead to results with the sort of collaborative behaviour that fosters improvement.

Data collection

Quality improvement requires the collection and analysis of data on patient needs, patient satisfaction and patient values and preferences, as well as data on outcomes. These are types of data with which doctors have had little experience. Measuring important variables within processes is often even more difficult than measuring outcomes.

Collaboration with patients

Patients have become used to a passive role, accepting the advice of doctors without inquiry. Doctors have assumed that they know the needs of the patients, or can judge what the patients wants better than can the patients themselves. TQM should broaden that role to ask not just, 'What can I do for you?' but also, afterwards, 'How well have I done it for you?' The health care team, including doctors will need training and support to learn to ask these questions.

Collaboration with management

Medical staff in hospitals spend an extraordinary amount of their time criticising managers. Managers complain about consultants who do not seem to be controllable, or willing to behave realistically given financial constraints. Little energy is left for collaboration. Care in modern medicine is complicated and crosses between clinical and management domains.

Collaboration is necessary if processes are to be improved. Such collaboration requires knowledge by each party of the work of the other. Both must be willing to respond and to change long-standing habits, assumptions and processes of work. Clinicians must be helped to understand and respect the many sciences of management. Managers need to acquire a working knowledge of the conditions under which doctors work and make decisions by regular visits to the wards and clinics. The most effective leaders have always led from the front line, where the action is.[27] For managers to collaborate effectively with doctors and nurses in management decisions, indeed to be able to communicate with health professionals, these skills are essential.

References

1. The Royal College of Surgeons of England. *Guidelines to clinical audit in surgical practice.* London: RCS, March 1989.
2. HMSO. *Government White Paper. Working for Patients.* London: HMSO, 1989.
3. DHSS. *United Kingdom reports on confidential enquiries into maternal deaths 1985-87 and 1989-90.* London: HMSO.
4. Royal College of General Practitioners. *What sort of doctor?* and *Assessing quality of care in general practice.* London: RCGP, 1985.
5. Buck N, Devlin HB, Lunn JN. *The report of a confidential enquiry into perioperative deaths.* London: The Nuffield Provincial Hospitals Trust and Kings Fund for Hospitals, 1987.
6. HMSO. *The quality of medical care. Report of Standing Medical Advisory Committee.* London: HMSO, 1990.
7. Gruer R, Gordon DS, Gunn AA, Ruckley CV. Audit of surgical audit. *Lancet* 1986; **1**: 23-6.
8. Bhopal RS, Thomson R. A form to help learn and teach about assessing medical audit papers. *BMJ* 1991; **303**: 1520-2.
9. Cookson C. Surgical route to efficiency. *Financial Times.* February 15 1989.
10. Shaw CD, Costain DW. Guidelines for Medical Audit: seven principles. *BMJ* 1989; **299**: 498-9.
11. HMSO. *Government White Paper. Working for Patients.* London: HMSO, 1989.
12. Shaw CD. *Guidelines for medical audit.* London: King's Fund Centre, February 1989.
13. The Royal College of Surgeons of England. *Response to the White Paper 'Working for Patients'.* London: RCS, April 1989.
14. Frater A, Spiby J. *Measured progress — medical audit for physicians. A manual of theory and practice.* North West Thames Regional Health Authority, 1990.
15. Bottomley V. Secretary of State for Health. Speech at a London Conference, April 1990.
16. Whates PD, Birzgalis AR, Irving M. Accuracy of hospital activity analysis codes. *BMJ* 1982; **284**: 1857–8.
17. Rees JL. Accuracy of hospital activity analysis data in estimating the incidence of proximal femoral fracture. *BMJ* 1982; **284**: 1856–7.
18. Crawshaw C, Moss JG. Accuracy of hospital activity analysis operation codes. *BMJ* 1982; **285**: 210.
19. Bennett J, Walshe K. Occurrence screening as a method of audit. *BMJ* 1990; 1248–51.

20. Harris CM, Heywood PL, Claydon D. *A guide to audit and research. The Analysis of Prescribing in General Practice*. London: Department of Health, 1990.
21. Difford J. Computer controlled repeat prescribing used to analyse drug management. *BMJ* 1984; **289**: 593–5.
22. Clarke K. Minister of Health. Speech, July 10 1990.
23. Shaw C. *Medical audit. A hospital handbook*. London: Kings Fund Centre, 1989.
24. Department of Health. *Working for Patients: Medical Audit, Working Paper No. 6*. London: HMSO, 1989.
25. Harman D, Martin G. *Medical audit and the manager*. Birmingham: Health Services Management Centre, **2**: 12.
26. The Royal College of Surgeons of England. *Guidelines to clinical audit in surgical practice*. London: RCS, March 1989.
27. Peters T. *Thriving on chaos. Handbook for a management revolution*. London: Pan Books, 1989, 423.
28. White T. Making Medical Audit Effective. Module 7, *Achieving Change through Audit*. Open University/Joint Centre for Education in Medicine, 1991.
29. Working for Patients Medical Audit Working Paper No. 6, Section 3.1, p. 5.

6

Managing change
Tony White

Introduction	96
Organisational change in hospitals	97
Types of change	98
Bringing about organisational change	100
Bringing about change	101
Change management	101
The steps in making a change	102
Including others in change	107
Handling change	107
Pathways and barriers to change	107
Commitment planning	110
Actions and awareness for change	111
Dealing with resistance to change	111
Evaluating change	111
References	112

Everywhere the old order changes and happy those who can change with it. (William Osler)

Introduction

There is very little use in asking a manager in industry for advice about change in medicine, except in the most general terms. The nature of medicine, the distribution of power and influence, the degree of external controls, the outlooks of the professions involved and, above all, the question of prime responsibility and clinical autonomy all interact to limit the changes that are possible.

Gale and Grant[1] state that 'advice must be firmly anchored in the context of medicine and must take account of its special nature'. They indicate some very useful guidelines on the best ways of managing change involving doctors, which will be of value to nurses and managers who are involved in any changes in medicine in its widest sense. When time is short, it is important that effort be used to best effect. Unfortunately we all tend to have limited amounts of energy and enthusiasm, and do not want to expend it on fruitless and unsatisfactory attempts at change. I hope to be able to show you how best to use the time you have available in order to produce changes effectively, which will be lasting and satisfying. It is important to choose the best route, but you have to know which route is the best in particular circumstances, and if you don't have, or have never seen, a map, this can be difficult or even impossible. It is important not to consider this advice as a construction plan or a blueprint of how to do things, but rather a review of a map of possible routes for successful change.

Organisational change in hospitals

Well-handled change can be stimulating, exciting and rewarding. Change that is handled badly can create anger, frustration and resentment and there may well be effects on quality and efficiency of patient care. Change is not easy, and never has been. Machiavelli[2] wrote in 1513 that:

> It should be borne in mind that there is nothing more difficult to handle, more doubtful of success, and more dangerous to carry through than initiating changes.

Interestingly he went on to say that:

> The innovator makes enemies of all those who prospered under the old order, and only lukewarm support is forthcoming from those who would prosper under the new. Their support is lukewarm partly from fear of their adversaries, who have the existing laws on their side, and partly because men are generally incredulous, never really trusting new things unless they have tested them by experience. In consequence, whenever those who oppose the changes can do so, they attack vigorously, and the defence made by others is only lukewarm.

Organisational changes, however, do have to be faced by medical, paramedical and other professionals, and also managers. That might involve changing methods or changing objectives, or both, and the demand for change has never been greater due to a current feeling that the system is not as efficient nor as effective as it might be.

Organisations, it is said, usually exist in a state of equilibrium, with numerous inbuilt forces and balances perpetuating themselves and rejecting changes that disturb the equilibrium.[3] Change may be forced on an organisation by external forces, although change can be made without this direct external pressure, when it is usually known as innovative organisational change. In spite of this rule there is no generally accepted definition of 'innovative organisational change'. Organisations that only change due to external pressure can, however, hardly be regarded as innovative.

Change in hospitals can be roughly divided into one of two categories: (i) external and (ii) internal.

External change

External change occurs when there are changes in the hospital's state of adaptation to the population and the district it serves. This reflects the way in which hospitals are continually struggling to adapt to a changing environment, over which they have only very limited control. Organisational change is the means by which a hospital is able to deal with the challenges imposed from outside, whether these are patient demand, general practitioner demand, government policy change, public demand, local pressures or technological advances.

In general, most hospitals tend to make organisational changes in reaction to external pressure. Occasionally they may make changes in anticipation of outside pressure but this is rare. It is those hospitals and units in the last group, i.e. that lead, rather than follow the pattern, that are attempting to change the environment as well as themselves. The creation of a new speciality department within a hospital, perhaps the treatment of infertility by *in vitro* fertilisation, or the introduction of a well established speciality into a hospital that previously had no such department, such as cardiothoracic surgery, would represent change due to external pressure.

Internal change

Internal change occurs when there are changes in the internal behaviour patterns of the staff. For a hospital's level of adaptation to improve, the behaviour patterns of at least some of the staff within that hospital have to modify, either their relationships with one another or to their jobs or both. One or more consultants may return from abroad or study leave with experiences of new ways of working. Here the development of new specialisations within the hospital evolves as a result of internal change. Audit may also identify deficiencies in the system leading to the implementation of change.

Types of change

Kaluzny and Hernandez identify three types of change as a function of whether ends or means or both are involved.[4]

Technical change

Technical changes do not represent any change in objectives or goals of the hospital, but involve some modification of activity due to new technology or a change in the structure of the organisation. For instance, a decision to install a CT scanner, MRI scanner or computer-based information technology system, although not changing the goals of the hospital, may have a major overall impact as it might well have marked finance implications on available resources. Changes in structure are sometimes more difficult to define than changes in technology but, for instance, organisational development or quality assurance programmes do not change the actual goals of the hospital.

Transitional change

Transitional change occurs where the organisation's goals change but not the means of achieving them. For example, the treatment of animals by radiotherapy in NHS hospitals or the sale of NHS hospital services for other purposes. These occur less frequently and are usually more stressful.

Transformational change

This is the most extreme form of change. Change occurs in the means and ends of the organisation. It is the least frequent and most unusual change. It might occur where a hospital has diversified traditional inpatient and outpatient services with services to fundholding GP practices held at the GP's practices rather than in the hospital, or maybe by the provision of shops and other facilities on the hospital site.

First and second order change

Changes are sometimes referred to as:

- First order change, where few systems and only a few people are involved. For instance, a department may change its way of working, perhaps deciding to perform all future operations of one particular type as day cases in future, or setting protocols for pre-operative investigations with the department. Because few people are involved the changes are relatively easy to introduce.
- Second order change, which often involves more of a cultural change. Many more people may be involved, perhaps several departments and therefore many interactions. A hospital may decide to increase very significantly its proportion of day surgery and this might involve surgeons, anaesthetists, junior doctors, nurses, general practitioners and managers, perhaps in the creation and organisation of a new day surgery unit.

Bringing about organisational change

You can look at *what* is to be changed or emphasise the process of *how* change is to be produced. The *what* may be considered under the headings of structure, technology and people.[5] The *how* changes are by approach and contained in Greiner's work which identifies seven frequently used managerial approaches.[6] These are:

Unilateral power

(1) The decree approach. This is a one way instruction from person in higher authority to someone in lower authority.
(2) The replacement approach. Where a key individual or individuals in an organisation are replaced in order to bring about a change. This can be seen when a new manager, and even a new Health Minister, is appointed.
(3) The structural approach. Where the basic organisational structure and its relationships are changed in order to bring about the required change in the organisation. The introduction of general management in 1983 is an example.

Shared power

(4) The group decision approach. Where the group members choose from several alternative solutions specified in advance from their superiors. It emphasises neither problem identification or problem solving, but the obtaining of agreement to a particular course of action.
(5) The group problem-solving approach. Here problem identification is shared and problem solving is through group discussion.

Delegated power

(6) Data discussion approach. Organisational members are encouraged to develop their own analyses of data presented as case material, survey findings and data reports.
(7) Sensitivity training approach. Where managers are trained to be sensitive to the underlying processes of individual and group behaviour. Changes in work patterns are assumed to follow changes in interpersonal relationships. Working on interpersonal relationships will, it is hoped, result in improvements in work performance.

In fact it is rare for one strategy alone to be adopted. A balanced approach using several methods is usually required. Successful change usually requires a sharing approach. The less successful approaches are those that use exclusively one or other of the extremes, set out above.

The management of change is a complex process and is helpfully thought of as a series of evolving stages. Lewin[7] identified three phases:

(1) Unfreezing. When people recognise the need for change or the *status quo* is jolted.
(2) Changing. Involving the introduction and implementation of change.
(3) Refreezing. Which reinforces the new patterns of behaviour.

Bringing about change

Unfortunately change is not always easy to achieve, especially when a whole department or other specialties are involved. How is the change to be managed? How do you ensure that everyone will be willing to become involved in the change and give it their full support? Fortunately, research has already been done which indicates how this might best be carried out in medical environments. We also need a 'change process' to ensure that the change is properly put in place. The following example shows us how to do this. It is based on the only model of change that has been specifically developed for use in the medical context and has been successfully applied in a number of demonstration projects.[8]

Change management

Guidelines for change or those contemplating change should allow for choosing a suitable route rather than being too prescriptive in the way of a blueprint for action. Sometimes it might be appropriate to omit certain steps, or to have different starting and finishing points. The following is, therefore, not a recipe but a checklist. Producing change is not always the stepwise logical process sometimes described in management textbooks. According the Gale and Grant[1] a 'model of medical change' involves the following core activities:

(1) Establish need.
(2) Power to act.
(3) Design the change.
(4) Consult.
(5) Publicise the change widely.
(6) Agree detailed plans.
(7) Implement.
(8) Provide support.
(9) Modifying plans.
(10) Evaluate outcome.

The steps in making a change

The step-by-step approach recommended above encourages the participation of all those likely to be affected. In medicine it is particularly important to have participation and involvement as these are two of the keys to successful change. However, additional intermediate steps need to be considered:

Recognition of need for change

First of all an event, for instance a financial crisis or the appointment of a new senior manager, is normally the catalyst that precipitates change. The hospital or someone within it has to be ready for change. And the change has to fit with the existing value system of the organisation. It is then necessary for an approach or combination of approaches set out above and below to convince others to produce change and to join the change process. The proposed change must also produce a new consensus to fight the inertia, reluctance and opposition to change within the organisation. It is because of this last factor, in particular, that a deliberate, structured approach to managing change should be adopted.

Identifying or recognising the need for change is a crucial step of the entire change process and usually involves a process of consultation or lobbying, although sometimes it might be circumstance or chance that identifies the need for change. It is vital that this crucial stimulus for change occurs, as it enables the move on to the next step, which is to make a diagnosis of the problem.

The climate of change towards day surgery at present creates an opportunity for change. The pressure is not unduly overt at present, but that pressure may become greater in the future, thereby creating a more obvious need to change.

Diagnosis of the problem

Always remember that problems do not present as givens; they do not present neatly packaged. Problems are described by Ackoff as being found in 'messes'.[9] In any group it is likely that the various members will each identify a difficult situation as a different problem. In trying to solve a long waiting list for surgery, the surgeons may identify the problem as not enough theatre time. The managers may identify it as poor organisation of theatre usage. The nurses may identify it as a shortage of theatre staff. They may all be right to some extent, but in reality these may only be initial indicators of a complicated interlocking 'mess' of problems.

Analysing the available information

The next step in resolving the 'problem' should be approached by asking three related questions:

(1) What is the real problem, i.e. what is the diagnosis as distinct from the symptoms?
(2) What must be changed to resolve the situation?
(3) What outcomes are to be expected on changing the situation and how might they be measured? This will ensure that any change has the effect that you planned.

Identification of solutions and alternatives

It is important to consider alternative methods and strategies for dealing with the problem(s). It will be necessary to take into account the effects not only on the hospital and its patients but also on those professionals and others working within the hospital who have to implement the change. The leadership style, the characteristics of the hospital and its group norms may all place limitations on the amount and direction of change. Consider all the alternative strategies available, and the needs of the people and the organisation.

The power to act

It is, of course, essential that you have the power to implement your proposed change. Your personal position in the local environment may be important. You may need to involve key people, harness committees and if you do not have enough authority it will be necessary for you to borrow power.

Design the change

If the change you envisage is feasible you will need to consider whether any resources that may be required are available. You will need to consider the timing and the timescale. Decide where there are barriers and pathways, and who may be winners or losers. In other words, ensure that the plan is practical.

Consult

Ensuring that your plan is practical is but one of the useful benefits from consultation. You will need to consult appropriately, exhibiting leadership and developing teamwork. Talking with, rather than to, people and explaining the proposals. Consulting with colleagues, other departmental members, other departments if they may be affected, managers and anyone who might be involved to ensure their support for making a change.

It is important to involve all the relevant persons in the planning stage, not just those that attend the meetings. It is also helpful to delegate main responsibilities to give all persons a feeling of ownership of the intended change.

Selection of method

Here the judgement of the leader is the key factor. The method must be that most likely to achieve the desired result, but it is important that those affected by the change take part in the choice. Ownership of the solution is vital.

It is here that the important decision-making role of the doctor/manager is crucial. There are four main models of decision making:

(1) The rational or classical model, where the primary decision-making criterion is a maximised outcome. Here the decision is usually taken by one person with one objective in mind, which can usually be written in quantitative terms. The decision problem is basically of choosing the best course of action.
(2) The organisational or neoclassical model, which has a satisficing outcome. Here decision making is regarded as so complex that only a limited number of aspects can be dealt with at any time. Not having all the alternatives available, a choice is made to be good enough to obtain the objectives. Uncertainty tends to be avoided by having short-term feedback to allow for changes in policy if necessary.
(3) The political model, which aims for an acceptable outcome. This is the exact opposite of the classical model, it has orientation to short term results in the same mould as the organisational model. The difference, however, is that the political model employs a compromise or bargaining strategy aimed at an outcome that is acceptable to as many external influences as possible. It focuses on differences, small numbers of alternatives, limited consequences and is not geared to long-range benefits.
(4) The process or managerial model, which has an objectives-based outcome. Similar to the classical model but with important differences, the process model is very much orientated to long-term results, has a definite planning mode and looks to the future, i.e. it is much more strategic in orientation.

Whatever the model used, decision making is often thought of as a choice between two or more alternatives and in this it is an essential part of the planning process and permeates all functions of managing. In fact, some authorities regard decision making as synonymous with managing. The decision-making process is, in many ways, a miniature change process (Table 6.1).

Table 6.1 Similarities between change processes and decision making

Change process	Decision making
Recognition of need for change	Recognition of need for decision
Diagnosis of the problem	Defining the problem
Analysing available information	Analysing available information
Identification solutions and alternatives	Identifying solutions and alternatives
The Power to act	
Design the change	
Consult	
Selection of method	Choosing an alternative
Publicise change widely	
Agree detailed plans	
Implement the change	Implement decision
Provide support	
Modify plans	
Evaluate outcome	Evaluate outcome

It is necessary to select the strategy that is most likely to achieve the desired result, although people must become involved in the change and participate in that choice; this is done in the next two steps.

Publicise change widely

Any planned change must be publicised widely and the presentation is very important. It may be necessary to amend proposals in the light of discussions with all interested parties. Through consultation, discussion and circulation of plans you insure that everyone knows about the proposals before the change takes place.

Agree detailed plans

It is important to agree detailed plans with all those involved and be sure that everyone agrees to try the detailed plans and understands their role within it. Only then can you consider moving on to the implementation of the change.

Implement the change

This is the next most crucial step and, if not well managed up to this point, can become the most difficult. At this stage one should stop thinking about change and actually make it — a managerial switch from process to task orientation. If people have realised from all that you have done before that change is necessary, they are more likely to make that adjustment. Sometimes change on a trial basis is more acceptable. This may help to reduce the feeling of insecurity that some people may experience. It may be helpful to implement the change in small increments, as this can also decrease the chance of the process going off course.

Provide support

You may meet resistance, which will need to be overcome; there will almost certainly be difficulties and objection. You can help by providing support to maintain the change process. You may understand the need to change the culture and this involves managing the stress that can accompany that change. Base support on existing systems and try not to destroy traditions or existing customs; practices and informal relationships should be disturbed as little as possible. Tolerate the stress and emotions of people, never meet hostility with hostility or emotion with logic and always allow resentment to be aired.[10]

Modify plans

It may be necessary to modify your plans. Be prepared to do this. Keep your eye firmly on the change process.

Evaluate outcome

Always evaluate the outcome to ensure that change has had the desired effect. Evaluation of the change is essential to ensure maximal utilisation of resources and to provide feedback, which may lead to necessary corrections or even the need for further change.

To summarise, the essential factors to achieving successful management of change are:

- Thorough consultation with all interested parties.
- Given that problems do not present as givens, ensure that everyone agrees that there is a problem that requires a solution. It is wise to present the problem, not the solution you have thought up.
- Give everyone a sense of ownership of both the problem and its possible solution.

- Ensure the practicability of the plans.
- Publicise discussion and plans openly.
- Ascertain the agreement of everyone to try a change.
- Provide support to everyone involved in the change.
- Be prepared to modify the plan.
- Evaluate the outcome.

Gale and Grant[8] give many practical suggestions for achieving each of these factors.

Including others in change

While it is usually not too difficult to change oneself, most medical practice today is based on the team approach, or what Freidson[11] has called colleague-dependent practice. In these circumstances, improving total clinical care for the patient usually involves others — colleagues, junior doctors, trainees, other professionals and managers — and is not just a matter of changing an individual's practice. Bringing about necessary change, which involves others, requires particular and distinctive actions. Peters[12] illustrates the 'astonishing effect of worker participation' with three examples of the power of involvement of others.

Handling change

Colleagues and managers should be involved from the very beginning, by letting them know what you are planning to change and keeping them informed of progress. If a problem is found, share that problem with these others and ask them to suggest possible solutions. Give people regular feedback. If the management is unable to allocate resources or the money then aim to involve them in suggesting alternative solutions. Always fix a date for a future review of the problem after recommendations have been accepted and implemented. It is important to ensure that action is being taken after the need for change has been identified. This is a form of procedure that will most certainly require the involvement of managers.[13]

Pathways and barriers to change

Not everyone will be enthusiastic about change and you will not necessarily find it easy persuading colleagues to change. It is helpful to ask why some are less than enthusiastic or even apprehensive about change. Address their worries. Give them the opportunity to work with their own solutions. It is also the best use of your time, as someone trying to implement change, to concentrate on those barriers to change that you have identified. It is always helpful for them to be aired openly and it has usefully been grouped under different headings.[13] For example:

Personal

- Exposure of weaknesses.
- Fear of derision.
- Fear of peer hierarchies.
- Loss of confidence.
- Loss of pride.
- Loss of status.

Interpersonal

- A not talking to B.
- C does not like D.
- E is jealous of F.

Intergroup

- Consultants' loss of status *vis à vis* managers.
- Doctors' distrust of managers.
- Managers' distrust of doctors.
- Rivalry between departments.
- Relationship difficulties between hospital doctors and general practitioners:
  — inadequacy;
  — pay differentials;
  — dependency on GPs for consultant practice;
  — limited understanding of each others problems;
  — limited opportunity to address these problems.
- Relationship difficulties between junior hospital doctors and consultants:
  — references;
  — hours of work;
  — quality of training;
  — quality/quantity of support of juniors by seniors.

Organisational

- Shortage of resources.
- Introduction of new management structure.

The existing pathways will usually look after themselves. A very useful mechanism for identifying what Gale and Grant[8] call 'pathways and barriers to change' is the force field analysis. This is just a way of listing all the factors that are in favour of the hoped for change and all those that are against it. You can then see which barriers you need to start dismantling or circumventing. The Templeton Series on DGMs[14] states that the:

> The more successful DGMs are those who have a well thought out plan or strategy based on a good understanding of the key people and the key forces for or against change.

Which is just another way of acknowledging the pathways and barriers. It is only when the driving forces for change are greater than the restraining forces that change can occur. And it is here that the force field analysis is useful for improving the effectiveness of change.

Examples of positive forces or pathways to change

- Adequate time for planning.
- Resources available for change.
- Pressure from patients general practitioners or colleagues to change, i.e. peer pressure.
- Encouragement from managers.
- Previous experience with good results.
- Training in change available.
- An enthusiastic team or department with positive help from colleagues.
- A personal interest.
- An approach from other groups.

Examples of some negative forces or barriers to change

- No time available to consider problems or solutions.
- No money for change (this may in fact be a stimulus).
- Need to change not agreed.
- No management support.
- No staff to help.
- No interest in training.
- Not personally very interested.
- Personal apprehension.
- Previous experience not good.
- Poor understanding of change process.

You will find that drawing up a force field analysis is a useful and effective tool to use at any stage of a planned or intended change. It can be used as a way of gauging the strength and quality of opposition to change as well as the forces for change. For change to occur the positive forces need to be stronger than those trying to preserve the current order. Change can be produced by higher forces, such as the introduction of the White Paper reforms, but it is usually necessary for an individual or small group up against great opposition for change to reduce that opposition rather than strengthen its arguments in favour of change. The dynamic nature of any organisation usually means that it is necessary to weaken the forces of opposition and there are a number of ways in which this might be achieved, for instance by:

(1) Encouraging key individuals to defect.
(2) Changing the design of change to avoid some of the opposition.
(3) Incorporating some of the ideas of those opposed to change.

Alternatively this analysis may help, by showing that the opposition to a particular change is too powerful. If you still feel that change is necessary or feasible it is sometimes helpful to put the forces in order of significance and tackle the strongest opposing forces first. Remember, that it is important that change minimises the number of people who lose position, status or influence, as they may become part of the opposition. It should also maximise those who will gain, as they could become supporters. Try to look for favourable local factors, pathways to change that can be a basis for success. And look for hurdles or difficulties that need to be overcome or avoided. Barriers are not absolute, they are related to what is proposed, how it is proposed, the content and style of change. If the style of change used is too directive or coercive the benefits of speed of change and integrity of original idea may be lost in resentment by some of those affected.

When talking to people it is important to present them with the problem and not the solution. Talking, explaining and sharing the problem involves people in the solution and gives them ownership of that solution. Lasting change can only be achieved by involving all those affected and accepting long timescales and maybe some dilution of the original concept.[8] One barrier to change may be a direct lack of involvement of other groups including other professional groups. The change process already described should indicate some ways of tackling this problem.[13]

Commitment planning

This is a useful way of assessing your support for change. There are four levels of commitment:[13]

(1) Opposers. Those likely to oppose any change.
(2) Uncommitted. Those who will not oppose any initiative but will not actively support it.

(3) Helpers or followers. Those who will support the change with time and other resources, provided someone else takes the lead.
(4) Leaders. Those who will lead the change process and make it happen.

Actions and awareness for change

You need to be aware of what changes are required. At the end of any meeting to achieve change, make an action plan:

(1) Document actions.
(2) Set priorities.
(3) List individual responsibilities.
(4) Identify a timescale.
(5) Agree attendance and date of next meeting.
(6) Agree who needs to be approached.
(7) Identify all other groups or individuals who may be affected.
(8) Make sure all plans for change are communicated.

Dealing with resistance to change

Dealing with resistance to change need not be a problem but must be considered before implementing any change. Resistance should be anticipated, or at least identified, as early as possible, as there is a natural human resistance to change, which may be very threatening to some people for many reasons. Unless it is recognised, the change may be diluted, delayed or destroyed.

Resistance may present in many ways and from a variety of directions, whether from colleagues, general practitioners, nurses, managers or even patients. It is often due to self-interest, misunderstanding and lack of trust. The stragegy for dealing with resistance has to be relevant to the problem but the types of strategies that can be employed include education, communication, involvement and participation, support, negotiation, manipulation and coercion, either explicit or implicit.

Each method may or may not be appropriate in a given situation and each will have advantages, some of which may or may not be ethically defensible in a certain situation.

Evaluating change

If you have handled your changes in the way described above, evaluating their effectiveness will be easier. The purpose of change will have been the need or perceived need for change or the need to exploit an opportunity for improvement. Once change has been made and the dust has settled it is always useful to evaluate the change to see whether it has met your desired objective(s). It is, of course, essential to take into account the viewpoints and feelings of all those involved in the change as well as those who lead the change.

Change may destroy existing organisational structures and the response of the organisation to the new equilibrium may not be that intended by the change leader. Change is an organic process. Many changes do not finish exactly where they were intended, although it may still be acceptable. Change badly handled can easily go wrong.

Change is difficult to bring about and requires considerable enthusiasm and energy to overcome the inertia resisting it, so much so that sometimes it may be difficult to predict the outcome with any degree of certainty. The outcome may differ what was originally envisaged or even desired. The change may even suit the majority but not the desired objectives of the original planner. The evaluation process requires a review of the original aims and objectives:[13]

- Change may have a long time span.
- Consult all the interested and involved parties as part of the evaluation of change.
- Be sure that you have improvement, not just change.
- Consider how practice has been changed, the resource implications and the impact of the change.
- After any change, consider whether it has led to the expected benefit and if not why not, and if it has, has it been maintained.

References

1. Gale J, Grant R. *Managing change in a medical context: guidelines for action*, p. 8. London: Joint Centre for Educational Research and Development in Medicine British Postgraduate Medical Federation, 1990.
2. Machiavelli N. *The prince. VI. New principles acquired by one's own arms and prowess*. Harmondsworth: Penguin Classics. Translated by George Bull, 1981.
3. Davis K, Newstrom JW. *Human Behaviour at Work: Organizational Behaviour*, 7th ed. New York, McGraw-Hill, 1985, p. 236.
4. Kaluzny AD, Hernandez SR. Organizational change and innovation. In: Shortell SM, Kaluzny AD, eds. *Health care management: a text in organizational theory and behaviour*, 2nd ed. New York: Wiley, 1988, pp. 380–1.
5. Leavitt HJ. Applied organization change in industry: Structural, technological and human approaches. In: Cooper WW, Leavitt HJ, Shelly MW III, eds. *New perspectives in organization research*. New York: Wiley, 1964.
6. Greiner LC. Patterns of organizational change. *Harvard Business Review* 1967; **May–June**.
7. Lewin K. *Field theory in social science*. New York: Harper & Row, 1951.
8. Gale R, Grant J. *Guidelines for change in postgraduate and continuing medical education*. London: Joint Centre for Education, Research and Development in Medicine, British Postgraduate Medical Federation, 1990.

9. Ackoff RL. *Redesigning the Future*. New York: Wiley, 1974.
10. Longest BB Jr. *Management practices for the health professional*, 4th ed. Norwalk, Connecticut: Appleton & Lange, 1990.
11. Freidson E. *Professional dominance*. Chicago: Aldine Publishing Co, 1970.
12. Peters T. *Thriving on chaos*. London: Pan Books/Macmillan, 1987, pp. 287–9.
13. White T. *Making medical audit effective. Module 7. Achieving change through audit*. Buckingham: Open University/Joint Centre for Education in Medicine.
14. Templeton Series on District General Managers. *Managing for better health. Issue Study No 5. Managing with doctors: working together?* NHS Training Authority. Templeton College. The Oxford Centre for Management Studies, p. 13.

7

Managing meetings
Tony White

Introduction	114
Types of meetings	115
General observations	116
Reasons for meetings	116
Some important issues of meetings	118
Some problems of meetings	119
Productive meetings	119
Why go to a meeting?	120
Who should go to meetings?	121
How many people should go to the meeting?	121
Committee meetings	122
Preparing for a meeting	122
Group dynamics	123
The theatre of meetings	124
The cast	124
Social and process leaders	126
The disguised agenda	126
Small group and brainstorming meetings	126
When to attend	127
Where to meet	127
Seating	127
Things that can go wrong	128
General meeting techniques	129
References	131

Littera scripta manet. (Latin proverb)

Introduction

Meetings are a central feature of professional life — a sort of organisational central nervous system. They are also expensive. It has been estimated that in a medium-sized hospital the cost of meetings might be in the region of a quarter of a million pounds per year.[1] Meetings are considered vital to an organisation as complex as the NHS and are a common event for nurse managers, business managers and clinicians representing colleagues. They are also an important management tool but many can be a waste of time and certainly most hospital meetings take twice as long as necessary. They account for unproductive time when:

- There is no clear cut objective for the meeting.
- There has been a lack of preparation.
- There is no agenda.
- There are too many participants.
- The wrong participants are involved.
- Ideas are not presented concisely.
- There is no leadership or control.
- There is improper use of visual aids.
- There are too many digressions and repetitions.
- Time is wasted on why, rather than how.
- There is a lack of clarity on outcome with unclear final decisions or even lack of decisions.

Meetings that do not achieve results are a total waste of time and only lead to more meetings. This chapter is concerned with answering these problems constructively.

Types of meetings

A meeting can be of two or more people and be anything from a small group talking together to a large public conference. There are many different kinds of meetings, with as many different names; small group, support, events, clinical, staff, departmental, open or public meetings, committees, workshops, learning sessions, conferences and so on. Every well-run meeting, however, whether formal or informal, is based on three essentials.

(1) Clear aims.
(2) Careful planning.
(3) Working together.

It is important that the aims are clearly agreed by everyone, the most successful meetings are usually those that have been carefully planned. Committee meetings need very careful planning. Everyone needs to work together and agree on the aims if not the detail. Clear aims will to some extent be related to the purpose of the meeting. A small group, whether this be a patient support group or a departmental meeting, may be for talking, listening and sharing problems. A committee, however, normally aims to make and agree decisions, planning how those decisions should be carried out and share out the tasks amongst its members. On the other hand, a learning session is about sharing ideas through teaching and learning. Of course, meetings may have more than one purpose and concentrate on a particular aspect during relevant parts of the meeting, so the membership has to be very clear about its purpose and when the style changes.

It might be useful to consider some characteristics of various hospital meetings:

(1) Routine and regular meetings, including committee meetings.
(2) Small group meetings arranged for a specific purpose.
(3) Workshops, impromptu meetings and brainstorming sessions.

General observations

You should always go to the meeting knowing what you want to achieve. Think about the meeting and never go unprepared. What decisions do you think will be made, what problems will be solved, what actions will be taken what information can be learned? Do not assume that what you wish to achieve will be recognised by others. Too many meetings are run on the assumption that everyone knows what the task is. Ensure that you and the others can define the objective of the meeting. If you know what it is you are trying to achieve you can anticipate moves that might be preventing this. What is the group being asked to achieve by meeting? It may be to offer advice, when the members think it is a decision. Members may be required to give factual information but think they are there to provide subjective advice. So advance preparation is vital (this applies to all members of the group) and the objective of the meeting must be clear to everyone.

Big time wasters

- Fighting losing or lost battles.
- Discussing items when the decision has already been made elsewhere. Especially when you are aware of this.
- Discussing items that are not within the group's power to decide.

Reasons for meetings

There are a number of ways of classifying meetings, such as those for planning and coordination, information, consultation, statutory, training, teaching, audit, quality, business, finance and so on. I am especially concerned with meetings involving clinicians, usually referred to as committees.

There are, of course, many reasons for holding meetings and I have tried to group these answers into a number of broad categories (with subdivisions, some of which may appear under more than one heading). They are applicable to most of the common reasons for holding meetings in hospitals:

(1) Decision making:
 (a) Organising.
 (b) Delegating.
(2) Discussion:
 (a) Socialising.
 (b) Consulting.
 (c) Persuading.
 (d) Generating ideas.
 (e) Motivating.
 (f) Activating.
 (g) Organising.
 (h) Isolating problems.
(3) Information:
 (a) Gossiping.
 (b) Reporting.
 (c) Communication.
(4) Supporting:
 (a) Socialising.

We are thus able to see that there is a scale of achievement for meetings.

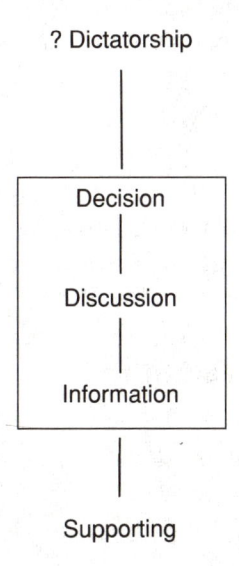

```
          ? Dictatorship
               |
        ┌──────────────┐
        │   Decision   │
        │      |       │
        │  Discussion  │
        │      |       │
        │  Information │
        └──────────────┘
               |
           Supporting
```

I have boxed those groups that are the most relevant to useful, productive committee meetings in hospital.

Some important issues of meetings

Socialising

The problem is usually that the members of the group are unaware of the purpose of the meeting.

Sharing work and responsibility

To share work and responsibility you must have small groups. It is an intricate area as, although it is usually easy to work or meet together, it is often difficult to do both.

Providing or receiving information

Meetings tend to be over-used to provide information and, while it may seem reasonable to clarify information, it should always be circulated in writing before the meeting. Failure to do this means inefficient and often ineffective meetings, or possibly that someone is trying to slip something through unnoticed. If you are looking for ways of reducing meetings or the length of the meeting, cut out those items that are for information only.

Persuasion and involvement

Make sure that everything that can be done before the meeting is done, the meeting can then concentrate on the primary objectives.

Creating and developing ideas

Brainstorming meetings are a means of providing the opportunity for the rapid creative development of ideas. They require frankness and freedom to make mistakes and should be conducted for the sole purpose of creating and developing ideas. It is also a good principle to make sure that all the ideas are on the table before critiquing them.

Delegation of work or authority

If the leader or a member of a group has not been able to define the objective of the meeting the meeting may end with nothing achieved.

Decision-making on tasks or issues

The greatest problem with decision-making meetings is understanding what part of the decision-making process the group is responsible for. It is also important to distinguish between:

- Agreement to a decision.
- Consultation prior to a decision.
- A decision requiring the agreement of more than one group.
- Enhanced commitment to a decision.
- Gathering information for a decision to be made elsewhere.
- Making a decision.

It is important to know whether all the preliminary steps to making a decision have been completed.

Some problems of meetings

Sometimes meetings are arranged and take place for the wrong reasons. There may be numerous hidden reasons for meetings, some related to specific personal agendas:

- As a substitute for work.
- Wanting to share the responsibility for a difficult problem. To avoid a tough decision or even making a decision. It is common to avoid decisions or responsibilities by holding a meeting.
- Meetings as networking channels, which all too easily become 'talking shops'.
- As a means of spending time with peers or those higher in the organisation, to find reassurance or attention, or to reflect power. This is only useful in maintaining relationships and is rarely productive.
- Instead of taking time to prepare a written report short enough to read, a meeting may be called. This is not only wasteful it can be dangerous as oral communication can be subject to more misunderstanding.
- Some committees get into a habit of meeting even though they may not have any real purpose, although they may publish a lengthy agenda.

Productive meetings

It is important to recognise that there is a subtle but none the less important difference between a stimulating discussion and a productive meeting. At the end of every meeting the objective should be restated and the results of the meeting summarised. Assignments should be restated and the follow up actions required.

For a meeting that you are attending, ask yourself:

- What is the purpose of this meeting?
- What should I accomplish by the end?
- How will I distinguish my success from failure?

After the meeting ask yourself:

- Was I correct about the task?
- Were the other participants clear about the task and objectives?
- Did I achieve the meeting's objective as stated in the agenda?
- If not, why not?
- Was the meeting a success or failure?
- What could be done to improve the next meeting?
- What was done that should be discontinued?
- What could the chairman do to improve the meeting?
- Could you have done without the meeting?
- If so how?

This will not only improve the effectiveness of meetings but improve participants' responsibility and involvement. Meetings so often drag on endlessly and are wasteful because of a lack of shared objectives. How often have you been at a presentation of a new idea and, when participants or members have been asked to comment nearly everyone at the meeting has suggestions for improvements? The reason for this is that they had not been given the opportunity to study the proposals in advance and were seeing them for the first time. Hence discussions are lengthy, suggestions are numerous and it can take a long time for the plan to be revised. If the committee members had been consulted beforehand, better results might have been achieved in a shorter period of time. So aim to get alliances and support in advance of meetings.

Why go to a meeting?

Any meeting you attend is your meeting too. Do you know why you're going? And remember that if you call a meeting you risk losing your meeting to a participant who is experienced in the management of meetings. So you could end up somewhere other than you intended. At every meeting you must have a personal objective(s). It may be the same as the stated objective of the meeting, or it may be different, in which case you may have a problem. You have to weigh the objectives and make choices. Your interest may be only marginally different or totally different from the others. In the latter case you are attending a different meeting from the rest of the group. You may feel the need to move the meeting to a conclusion that will only benefit yourself and that may not be in the hospital's best interest. Don't forget that personal agendas can destroy meetings, they do exist and members may be heavily defensive to protect their personal agenda. There may also be as many objectives as participants. So think strategically about the meeting.

When you have decided the value of having a meeting always reconsider the alternatives. Could you achieve your objectives with a phone call with potential participants, perhaps a letter would produce the necessary response, or maybe the purpose is only the dissemination of information when a sheet of paper would do as well? Satisficing is all too common.[2] In choosing, human beings do not consider all alternatives and pick the action with the best consequences. Instead, they find a course of action that is good enough — that satisfies. Organisations are happy to find any needle in a haystack, rather than searching for the sharpest needle in the haystack.

Who should go to meetings?

Only those who can influence the fulfilment of the objective should attend a meeting. Unnecessary participants, like unnecessary meetings, are again a waste of time. That does not mean that anyone who may oppose the objectives should be excluded; that may do more harm. Opposition can be overcome by including opponents in the discussion, at least it shows respect for their position, however much you disagree.

You might ask the following questions:

- Who is one obliged to invite?
- Who can give you what you want?
- Who is in favour of the objective?
- Who will oppose the objective?
- Who will sit on the fence?
- Who can cause trouble if not invited?

How many people should go the meeting?

It has been said that the length of a meeting is proportional to the number of people attending. However, a successful meeting is also dependent on who is present, and that depends on who is invited. The more participants, the more difficult it is to achieve objectives. Of course, hospital policy often determines who should attend, however, the more specific the objectives the fewer the participants should be.

The psychology of group dynamics is important. A meeting of two people should provide excellent communication and a useful outcome. Unfortunately the process may have to be repeated with several individuals and, if the outcomes are different, this could create problems. Small groups from three to nine individuals invite candour, intimacy and real results. Some people[3] '... are often reluctant to speak up in meetings of any sort involving more than four or five people, especially when they are juniors amongst a number of more senior people'. When the group reaches ten or more people, acting and a sense of theatre comes into play as opposing parties try to impress colleagues by aiming for effect rather than results. When the group reaches twelve or more then two leaders or groups are likely to emerge.

No one should feel the need to be at a meeting if the objective is of no particular concern to them.

Committee meetings

Some meetings are held on a regular basis. Routine committee meetings, like all meetings, are also only effective as long as they have objectives that can be fulfilled. Many hospital committees have long since forgotten their original objectives. Committees should regularly audit their value and purpose, as they rarely consider their function and usefulness, often forming new ones, in preference to disbanding old ones. The average British hospital has about forty committees, with approximately seven to eight consultants on each, mostly meeting at least monthly. Meetings held on a routine basis, weekly, monthly, etc. are mostly tedious and time-wasting. Situations are often created to provide topics for the agenda. Better to wait for a situation or problem that requires a meeting and then call it. Many larger meetings are a ritual to keep everyone up to date on what is happening. Each department is represented and although they may have had nothing new to discuss at their own divisional meeting they often produce lengthy minutes, which are raised with little relevance for other departments and nothing is accomplished. A simple report of each meeting could be circulated to other departments instead.

Preparing for a meeting

There should be concise accurate summaries with respect to the problems being addressed, with options already identified. If more information is wanted it can then be made available with the documentation, but not as part of the summary. Do make sure you are fully prepared for the meeting, otherwise you are failing to be an effective member of the team. Once the meeting is in progress you are at a disadvantage by your lack of preparation.

A meeting may be the opportunity to put across a message, so if you have an objective, always have a prepared statement available when the opportunity presents itself to state your personal objective. To make sure you don't miss a point, use notes but make it a specific concise message. Memorised statements sound artificial and the material controls you:

- Never read a prepared statement aloud.
- Never use a memorised statement.

- Speak naturally using notes.

A seed sown today may be important for later. You can often turn someone else's question into a bridge for your own message, even though the question might not be specifically addressed to you. A question in any situation will always give an opportunity to say what you want to say, a technique absolutely made for meetings. Always answer the question and then make your point. If you are asked a question and you don't know the answer it is perfectly acceptable to say 'I don't know but I'll find out and let you know ...' and then go on to your own statement, via a suitable link. You can also use the same technique in response to a statement by agreeing with that statement, before linking to your own statement. At this point the speaker becomes the leader and controls the meeting while he is speaking, and so long as the subject of the discussion continues he remains in control. This is very easy to observe at medical committees. It has been stated that 'what is said cannot be unsaid and it may stick in the mind of the listeners'. Unfortunately spoken words are also like smoke in the wind and disappear, so you need people to see your words mentally. Mental pictures or stories do that better. Often it is not the message, but how it is said that counts.

Group dynamics

It has been said[4] that 'there is a tendency for meetings to drift towards collective incompetence' and groups of individuals are far more likely to err than individuals. Regular staff meetings with a long agenda produce few results. Often, the same issues appear time and again for discussion as continued minutes. At the following meeting, people delegated assignments have not done them and participants find themselves saying 'why do I bother to come?'. However, it is possible to manipulate the sources of collective incompetence to achieve certain goals. At the heart of collective incompetence is the so called 'group mind' where people meeting together have an enlarged mind with tremendous capacity, which can achieve effective results but also, alas, has severe limitations. In addition you are not only dealing with memory in the group mind, but power, ego, emotions, ambitions, hidden motivations and many other factors more powerful than logic. So the group mind has a number of flaws. People dislike inconsistency and attempt to eliminate it. This mental conflict or cognitive dissonance occurs when information challenges group beliefs. The group reacts to preserve its preconceptions. This is common amongst medical committees.

Dissociation from the task may occur. A member may believe they are less well informed than the rest of the group and, although disagreeing, may vote in favour to remain accepted within the group. The group may reach a decision based on the lowest common denominator. This may not solve the problem but it saves feelings and egos. The factors affecting group thinking may be miscommunication, outside pressures, personal agendas and a pattern of meeting that becomes traditional and ingrained. Professor Cyril Northcote Parkinson is reputed to have stated that the time spent on any item on the agenda will be in inverse proportion to the sum involved.

Then there is the problem of strains between the various members, representing different groups, between their representational role and their corporate role, between their personalities and backgrounds. They may all have different visions of the future. They all require emotional maturity to face these tensions, but how rarely those tensions are allowed to surface. Everyone remains polite and reserved. It is unnatural for people not not talk about these discomforts openly. Quality does not emerge by talking about it. Teamwork does not result from talking about it. The group needs to do these things.

The theatre of meetings

All your actions communicate something to others, how you sit, where you meet, how you dress, the tone of your voice, your facial expression, the energy you create. All convey a message.

People listen and contribute more effectively when their attention is engaged, but holding their attention is difficult. Facts alone are seldom interesting, they need some degree of entertainment. Consider this before the meeting and perhaps plan a change of tempo. Show your interest, enthusiasm is infectious, remember theatre often works subliminally, and you can control without the appearance of control. A useful technique at meetings is the well known 'good cop, bad cop' routine. The good appears conciliatory and reasonable, the bad irrational. In reality they are on the same side.

Always consider the audience other than the one physically present at the meeting. Every person at the meeting reflects his other audience, sometimes directly and sometimes with out admitting it. One needs to learn to look for these audiences and be aware of the needs of people in serving them.

The cast

Although some seem to be heroes and others villains they may simply be responding to others at the meeting, the issues under discussions and in a manner consistent with their personality and needs. The following will not be at every meeting and frequently you can see more than one character in the same person.

First the Good Guys, who you need, but have to recognise for their role:

- The visionary. Has initiative and imagination, can get things started, can offer ideas and solutions. Visionaries often have large egos, which require recognition, so look to them for ideas but give credit and involve early. They will not be too interested in practical detail, being basically ideas people they respond to a challenge.
- The steady influence. Can re-focus the attention of the group when they drift. Although not so creative as the Visionary they are less likely to be taken off track with the excitement of new ideas. They fail to explore detail but it is useful to approach them before the meeting with what you want to accomplish. They can be your navigator keeping things moving on track.
- The interpreter. Older, clarifies without offending, ask questions, distinguishes arguments, can interpret and restate the group's position. Look to them when there is deadlock. They are natural administrators and can often provide the details.
- The smoother. Like the Interpreter, older and wiser. They have been around and seen it all before, and can reduce tension with an amusing comment. They do not want to be involved in the work but are happy to give advice, which is usually sound.
- The supporter. A supportive personality, encouraging, but not a leader, can find what is positive, but has difficulty with personal choices. They know what is right deep down, however, and will back you once the choice has been made. Do not put too much on them, they will probably decline to accept anyway.

And the Bad Guys you could do without but who are always around, you just need to recognise them:

- The aggressor. Questions everything. Wants attention. Sees problems but seldom offers solutions, seat them next to an ally if possible.
- The show-off. Shows obvious disinterest, engages in side discussions and reads other material. Attempts to remain uninvolved. Usually harmless and only wants attention.
- The know-all. Manipulative, seeks control, sometimes quite well informed but often not. Best controlled by seeking their advice prior to the meeting.
- Scapegoats. Scapegoats can be vital to meetings, and may not necessarily be present or even a member of the group or committee. They are often seen as saboteurs that can appear to block change, thereby providing the rest of the meeting with a reason for retaining the 'status quo' and avoiding any decision. As scapegoats are seen as bad guys, the rest of the meeting can then see themselves as good guys.

There are various classifications of the sort of people you need to make an effective committee or team but they all make necessary contributions to the task process. The best teams are not necessarily the brightest but those of the right mix.

Social and process leaders

This is a division useful to distinguish:

- Social leaders, who deal in terms of people rather than issues or process. They converse before and after the meeting and are sensitive to the feelings and emotions of other people enjoying the contact of meetings.
- Process leaders, who look for order, distinguishing process from structure. They are very quick to focus and to notice when the discussion goes off track although they may not have the personal skills to retrieve it.

Meetings usually require both social and process leadership. It is useful to assess yourself and learn to fill in the gaps.

The disguised agenda

A document submitted to a group may not look like an agenda but might contain items that, if highlighted in the manner of an agenda, might engender debate. It is possible to rig an agenda, if one is not presented, by circulating a course of action that then becomes the agenda. If you want to get something on an agenda try and get it on early enough for a discussion.

Small group and brainstorming meetings

Usually a group of four to six people, always picked and assigned to achieve a specific task. The members are picked for what they can contribute. To set about the task it is useful to define how it is to be achieved, set tasks for individuals, the group can then meet to pool the information obtained. The process may need to be repeated until the objective has been achieved. When the task has been completed, however, the group should be dissolved. Failure to dissolve a group after it has achieved its objective results in yet another committee with no defined objectives absorbing resources.

When to attend

If you are invited to a meeting only attend if it will be useful to you.

Where to meet

There are a number of advantages to holding a meeting in your own office:

- You obviously exercise some degree of control, which would be impossible elsewhere.
- You control access to the phone.
- You will probably feel more comfortable.
- You have the opportunity to play the gracious host.
- It may save you time if the alternative is a meeting in another building.
- You can terminate the meeting when you choose.

Never crowd people into an office. If you arranged the meeting this is your responsibility. You are making a statement about your ability to manage even if you assigned the task to someone else.

Never insist on meeting in your own office if this creates tension, as there can be advantages to meeting on neutral ground. By deferring to the other person, you can gain an advantage.

Seating

Seating can have a profound effect on behaviour in a meeting. Sitting behind a desk creates a barrier. A peer should be met in a more comfortable area or around a table. You are more likely to get candid advice or new ideas when the person you are meeting is not feeling vulnerable.

If you are dealing with a peer in his office and who remains behind his desk you can diminish his control by remaining standing or moving about the room. You have a height advantage and his eyes have to follow you. You will soon find him away from the desk.

In face-to-face meetings the mood will be intensified, especially in large groups where the energy is focused. A full circle is most intense. This is not necessarily good or bad as such but depends on what you are trying to achieve. It does make control more difficult, as all the parties are of equal status, but discussions can be creative and intense. Semicircles are good for problem-solving meetings and provide a good balance of control and sensitivity. Everyone can see everyone else but they can also focus on you as discussion leader.

Long tables are wonderful control mechanisms but do have limitations. They are useless for brainstorming sessions. It is impossible to see others down the same side and as a result discussion between them is limited. This can be an advantage with two allied trouble-makers who, if placed a few seats apart on the same side of the table where communication is difficult, can have their effectiveness minimised. Obviously you will need to establish assigned seating to accomplish this goal. In informal meetings, however, you might consider where you might place the most critical participants, perhaps a valuable team player and a trouble-maker, planning where you want them in relation to you and make that a casual suggestion. Sometimes a group that meets regularly may establish tacitly assigned seating, which this can be observed at many regular meetings.

Always be aware of other sources of influence at a meeting and position yourself where you can easily make eye contact. The more eye contact and with more people, the more control you have. To some extent where you sit will depend on your purpose. If you want to be uninvolved pick a position that permits that. If you are seeking to win a point or plan seizing control, pick a controlling position. At a long table this will be either end. In a three-sided arrangement this will be either side of the chair or at either end. At a long table an ally at the other end of the table will give you maximum support in handling a difficult meeting. Next time you are at a meeting observe where people sit and see how it influences their roles. Observe the quiet, apparently unassuming member who has developed a technique to suit their own agenda.

The ideal position is opposite the chairman so that you can talk directly to him and include others as well. If you are going to be a competitor to the chairman sit as far away as possible. Where the participants sit may shorten the meeting when clear lines of authority are established and the meeting is kept under control.

There are two other useful seating points to remember. First, the comfort of the chair is inversely proportional to your energy level and that is proportional to the speed of obtaining your objective. Second, the more uncomfortable the chair the less likely you are to doze off. If necessary sit on the edge of the chair, this keeps you alert and adds energy, intensity and power to any presentation.

Things that can go wrong

No natural leader may emerge so that there is a general discussion with no solution to the problem, or an over-dominant person takes control, again with no decision. Alternatively there may be two strong members, which leads to antagonism, with the rest of the group becoming an audience and still no decision reached. Make sure there is communication outwards to others and that the group does not become a clique, secret gathering or cabal. And of course the group must dissolve itself after completing its task.

Highly effective meetings are:

- Task-orientated.
- Time-limited.
- Consist of people relevant to task.

General meeting techniques

Interruptions

When to interrupt to gain control is a meeting tool.

Method 1

'Just a second, may I . . .' and continue speaking.

Method 2

Call the person by name 'Fred, there's something else . . .' and continue speaking.

Method 3

Just start speaking and raise your voice above the level of the other person.

One thing you must not do, however, is wait for permission to be granted before you interrupt, but to continue talking.
    It is often difficult to hang back while someone presents something that you consider inappropriate. Do think twice before making yourself vulnerable by interrupting the speaker, your own point may not be quite as perfect as you thought. The chairman should remember that interruptions can lead to digressions, which can prolong a meeting, but this is dealt with in Chapter 8. Although a timely and polite interruption can cut off a digression. If you want to prevent yourself being interrupted,

Method 1

You may insist on finishing your point. Say 'Please, just let me finish . . .' and continue.

Method 2

Hold your hand up, palm outwards and continue.

If you should be the subject of destructive criticism. Ask the critic to present an analysis of your proposal with workable alternatives for the next meeting. Emotion can resist logic, you will not convince someone with logic when his mind is blocked with emotion. To overcome emotion you need to talk not criticise. When up against someone who is emotionally aroused your best ploy is to remain silent until their emotion peters out.

Opposition and confrontation

Try 'Your perspective seems to be . . . whereas mine is . . .'. When you are up against opposition your objective should always be to determine what your opponent's objectives are. There is nothing wrong with having an objective that is different from the leader of the meeting or other participants but you have to ask yourself what your opponent wants and why he wants it. Only then can you plan your own strategy. What are the strong points in favour of your opponent's objective, and the strongest points against? Who is on the side of your opponent and who are your allies? What are the strongest points that will enable you to accomplish your objective. You will see here that there are similarities to managing change as I described in Chapter 6. You must plan your strategy to increase your chances of success.

If you are disagreeing with someone try to provide your opponent with a way out, a way to save face. If you disagree, first state what you agree about, support their position as modified by your own. Make criticism less personal by claiming you are acting as the 'devil's advocate'. Meetings are pressure situations. It may be time pressure or peer pressure, everyone wants to look good in front of colleagues, not forgetting the problem of decision-making. You need to say what you want to say in the shortest possible time.

Contribute early

Research has indicated that when a person contributes early in a discussion he is more likely to exert a greater influence throughout the discussion. This forces opponents to respond to you, but be prepared to come back into the discussion to combat opposing points when you rebut their counter arguments.

Defending a weak case

When your case is weak consider additional techniques. Use the psychology of the seating positions. Use guilt by association as you associate your opposer's idea with a word or phrase that has negative connotations. Use glittering generalities that associate positive images with your idea. Use transference, where you buttress your idea with reference to respected authorities, and testimonial, where, like transference, you link your idea with respected individuals. Relate to the group suggesting, 'we are all in this together'. Offer your idea to the others in the meeting, sharing points in common and past successes. Cite only the strong points of the argument for your position and the weak points of the arguments against your position, which are easily dismissed. Suggest a bandwagon effect, pointing out that most people are in favour of your position. And finally use props, charts and graphs to divert attention away from the substance of your idea. The group may vote on the presentation rather than the substance. When you are in a winning position do not waste everyone's time continuing the argument, you have nothing more to gain but could lose, if the issue drags on.

Dealing with reluctant individuals

Try involving reluctant participants in a small group project and, if all else fails, build a wall around them to isolate the problem. Remember the bishop, when asked about recalcitrant priests. Death is a great resolver of problems; the same applies to retirement.

References

1. Knibbs J, Sellick R. Tell them I'm in a meeting. *The Health Service Journal* 1991; 11th April.
2. Allison GT. *Essence of decision. Explaining the Cuban Missile Crisis.* Boston: Little, Brown and Company, 1971.
3. Nelson MJ. *Managing health professionals.* London: Chapman and Hall, 1989.
4. Kieffer GD. *The strategy of meetings.* London: Judy Piatkus (Publishers) Ltd, 1988.

8

Managing the Chair
Tony White

Introduction	132
General observations	133
Characteristics of the Chair	133
The Chair's neutrality	134
Seating	134
Time	135
Control	135
Planning a meeting	136
Preparation for the meeting	140
Distinguishing success from failure	141
Standards for measuring meetings	142
Chair manners	142
Running the meeting	143
Dealing with problems	143
When not to hold a meeting	144

Experience is the name everyone gives to their mistakes.
(Oscar Wilde)

Introduction

It requires skill to manage an effective meeting and most doctors have had no training for this role. You can often predict the outcome of any meeting by who will be chairing it.

Control and authority come from the Chair's perceived commitment to the meeting and it's objectives, and the Chair's skill in supporting the group achieve those objectives. If you, as Chair, are thought to have your own agenda or support some members over others you can restrict your own authority. Your comments and objectives should be limited. The less you say, the greater your strength. Your primary concern is the interest of the group, the integrity of the meeting, the achievement of the group objectives and the aims and objectives of the organisation as a whole.

General observations

It is important to make any meeting a positive experience so that people are clear about their work, the aims of the hospital and its goals. It is all too easy for a meeting to become the centre of disharmony, especially if the group is large and the meeting goes on too long. Individuals become inattentive and bored, pomposity reigns, personal prestige becomes an issue and the activities, as well as some members, become defensive. The meeting becomes self-serving and disappointing. This creates the atmosphere that it is a terrible place to work in, full of problems and ultimately work and morale suffers. When a meeting bogs down the participants have probably lost sight of the objective. Many doctors never know what sort of meeting they are in, and meetings to inform a decision which has already been made will still debate it.

Characteristics of the Chair

The quality of the Chair is as much art as science, usually with the following characteristics:

Communication

One of the most important functions as Chair is to facilitate communication. You are the principle voice of the group mind. You must look out for possible misunderstandings, words or phrases whose meanings are unclear or ambiguous, generalisations that lose meaning. Watch for erroneous assumptions being made, statements removed from context and for a committee to assume that the last statement is its final position.

Anticipation

Visualise the meeting before it takes place. Organise the meeting with the tasks of the committee in mind. Ask yourself what objectives you are aiming to achieve from the meeting. Conceptualise the specific measurable objectives, the strategy and steps you will need to achieve them. Then make sure the group shares that with you.

Personnel management

Managing a meeting is about managing people, having an agreeable personality is not enough. You need to be able to sense dissension or harmony, agreement and opposition, confusion and certainty, and bring it all together. You need to unite the group and be united with the group. Focus on uniting rather than dividing the group. Be friendly, offer compliments. Although distance can confer authority it can also produce both concious and unconcious resentment. Leadership skills require you to accept people as they are not as they were, trust people and be able to do without constant approval and recognition in your role as Chair.

Adaptability

Although you should stay strictly with the agenda, be prepared to adapt, should something unexpectedly turn out to be important. This does not mean allowing digressions, repetitions and pontifications, and all the other wanderings one sees so often in medical meetings, but knowing when the discussion should be allowed to flow unhindered to achieve the objectives of the meeting.

The Chair's neutrality

The Chair represents the group to the group. Over 90 per cent of what you say should be a reflection of the group or the process, rather than you as an individual.

Decide your position before the meeting on all the major issues, although be prepared to change your mind. Make an effort to form preliminary opinions on all issues. Your opinion may be decisive, although during the meeting you must remain neutral. If you as Chair argue your position with peers you will have difficulty paying attention to the process, thereby placing your own identity over that of the group. Your attention should be directed entirely to the process and not the structure. You must also try to predict issues that require further discussion, or limited discussion and debate in order to structure the meeting effectively. Even deciding whether certain matters are in fact worthy of the meeting.

How and when you can influence the substance or structure should be rare, but you do determine what appears on the agenda and this is generally more important than what is said. If you want to influence structure it must be done subtly and sparingly. Only when all other efforts fail should you attempt to influence the outcome, by expressing substantive opinions. If you wish a particular point to be advocated it is far better to arrange for another member to do it for you. In the end your continuing effectiveness as Chair, as strong neutral and fair, the embodiment of the group as a whole is more important than your effectiveness on any one issue.

Seating

The Chair should sit in the most visible spot. From there he can be seen and heard and by seeing and hearing he is in a position to control the meeting. At the end of a rectangular table and farthest from the entrance is best. Or you may sit at the centre of the longer side of a rectangular table, but this does not command as much attention. Do not sit next to the head of the table as this gives authority to the person at the head of the table thereby weakening your own authority. If the table is round or oval visualise it as rectangular, and sit at the end farthest from the entrance.

Time

Remember that starting on time and ending on time is for the mutual benefit of the group. To make a dull meeting less dull shorten it by setting a time limit on it. And stick to it. Always state the time the meeting will end. If it does overrun it is invariably the fault of the Chair.

Where appropriate consider staggered attendance, minimising the amount of time each participant has to spend at a meeting. To have people turn up on time consider setting an offbeat starting time. Time management experts have shown that people are more likely to turn up on time for a meeting scheduled at 9.10 a.m. than one at 9.00 a.m. because the former is more specific, and many people allow themselves to be ten minutes late for a start on the hour.

If you want to shorten a decision-making process — vote. Make the decision even if you do not have every single known fact. People do not like decisions. People feel that no decision will keep them out of trouble. When discussion is endless you must vote.

Control

By this I do not not mean controlling directly what people say or how an issue will be decided, I mean the effective control of some of the group so that everyone can have a say to achieve the best decision. You have to gain control early, before the start of the meeting.

The first step is to avoid a meeting that is likely to go nowhere:

- More mediocre meetings mean less capable management.
- Fewer but better managed meetings mean more effective management.

The agenda is where the control of a meeting of peers originates. Don't neglect consultation, your leadership comes not from just creating the agenda but in consulting with your peers in shaping the agenda. Members assume that you alone possess the input from all other members and will shift to a quasi-subordinate role when assuming that you possess the whole picture. If it becomes clear that you have not consulted, then your control and credibility are justifiably threatened.

Your consultation also gives you the opportunity to reach any hidden agenda. You will then have more information than your meeting partners and this will assist you in achieving your aim from the group. Nothing breeds authority like success, a pattern of successful meetings, will have everyone wanting to work with a winner. Give credit where it is deserved otherwise you risk jealousies and loss of authority. It is always more difficult to move from informality to formality than vice versa so always begin on a formal note. In meetings with peers, subordinates and superiors if you want honest input make sure you the ask lowest ranking members first, however, if you want conformity ask the highest ranking person first.

To get the most out of a meeting, limit the risks and focus the group to achieve your purpose:

- Spend more time in preparation and less in the meeting.
- Limit the number and kind of tasks to be undertaken.
- Limit the number of participants.
- Start once all the vital participants are present.
- Keep the meeting on course by discussing only one issue at a time.

Spending more time in preparation and less in the meeting is important because meetings fail in inverse proportion to preparation and in direct proportion to meeting time. As a general rule, try to reduce the size of meetings. Limit the number of participants and the number and types of tasks to be undertaken, as meetings can fail in direct proportion to the number and variety of tasks. Once all the vital participants are there start the meeting, do not wait for latecomers. Keep the meeting on course by discussing only one issue at a time. Wider participation in meetings maybe healthy but can simply lead to the need for the creation of smaller subgroups to serve the function of the original group. For instance, some Medical Executive Committees, Medical Management Groups or Management Boards of Clinical Directors set up for policy setting have, with the emergence of more and more faculty and subspeciality interest groups, become so expanded with extra members that the committee reaches such a size that effective planning and control has become impossible.

Planning a meeting

He who controls the agenda controls the meeting because a skilled Chair does not allow anything on to the agenda that does not serve his own interests. Each meeting should have an agenda as a road map rather than a shopping list. A construction plan or blueprint for the meeting. Do not confuse a brief agenda with a brief meeting. Agendas are sadly often too brief and too vague, which only encourages people to pile on additional items. You must also tell the members of the meeting enough about the item for their contribution to be meaningful so summarise on the agenda the task and attach all relevant papers. Each item should provide enough information to allow the participant to do any necessary preparation and to understand what you hope to achieve through consideration of the item. There are also different kinds of tasks. You must indicate whether an item is for information, discussion, action, or all three. Meetings often fail because participants are not clear about what is on the table and what is expected of them. Circulate the agenda really early to allow members to do their homework. For regular routine meetings this is vital, to allow them the opportunity to seek advice and counsel of their own staff or colleagues.

It can be infuriating to see people at meetings, thumbing through sheaves of papers, all of which have been carefully prepared, trying to understand their purpose. How can such people hope to function as policy and decision makers? There should be concise, accurate summaries with respect to the problem being addressed, with options already identified. If more information is wanted it should be available within the documentation, not as part of the summary. Papers distributed at a meeting are of little use. Passing out a report at a meeting is a total waste of time and cannot be considered seriously, nor will it be. Neither will it be read afterwards. It is only valuable as a technique for bluffing an item through and is a distraction, shifting the focus away from the matter in hand. It is pointless to bring items without preparation and routine items often waste time with trivial discussion. It is sometimes helpful to schedule items in order of relative importance.

There are four roles for an agenda:

(1) A tool helping to prepare for the meeting.
(2) Telling the participants in advance what items are to be considered.
(3) A script for the meeting further symbolising the theatre analogy and thereby a mechanism for order and control (see Chapter 7).
(4) The standard by which success or failure can be measured.

The agenda should be the blueprint and plan of action for the meeting. It may, however, be necessary to request action plans from key players before the meeting, which can be discussed ahead of the meeting in order to produce an appropriate agenda. This will focus you and the others on the objectives of the meeting. This should include the objectives of the meeting, the issues to be discussed, the time when the meeting will begin and end, the place, the participants and what is expected of them by way of preparation before the meeting.

I have seen a slip of paper circulated from a Chair:

> There will be a meeting of the Medical Committee on Thursday 17th October at 8.00 p.m. in the Committee Room.

An agenda was then be distributed at the meeting, two hours were wasted in discussion on the various items, but nothing was decided and very little achieved. Everyone went away saying what a waste of time it had been.

Another hospital circulated the following:

There will be a meeting of the Medical Executive Committee on Friday 8th June at 6.00 p.m. in the Board Room.

AGENDA

1. Apologies
2. Minutes of last meeting
3. Matters arising
4. Meeting with Regional Medical Officers
5. Continued minutes
6. Financial report
7. Medical manpower
8. Clinical directors
9. Medical audit
10. Total quality management
11. District medical committee
12. Management of head injury
13. Divisional matters
14. Date of next meeting
15. Any other business

Under Item 5, continued minutes, three items were brought up for discussion lasting 15 minutes. No one had known in advance what they were, the Chair had apparently made no approach prior to the meeting to have the problems resolved so no decision could be made and the Chair was obliged to put them forward for further discussions at the next meeting.

Under Item 4, Meeting with Regional Medical Officers, no one knew in advance why these officers were coming. They were introduced and the committee was asked if they would like to ask any questions. After some silence and four brief questions there was silence again.

Item 6, the Financial Report, consisted of a number of papers presented at the start of the meeting, no one had therefore had a chance to read them nor time to consider and consult with colleagues.

Medical Manpower, Item 7, was just a list of consultants recently appointed, posts advertised and new posts being passed on for approval. In fact over half the items on the agenda were presented at the meeting with no prior information. Added to which, the Agenda only arrived with members the day before the meeting took place.

The agenda should have looked more like this:

AGENDA

1. Apologies
   For information
2. Minutes of last meeting
   Attached
3. Matters arising
4. Meeting with Regional Officers
   Regional Medical Officers attending to answer any questions from committee.
5. Continued minutes
   List itemised
6. Financial report
   See Appendix A. For information
7. Medical manpower
   Present situation as per attached document. See Appendix B. For information.
8. Clinical Directors
   General Managers' proposals for remuneration. See Appendix C. For discussion.
9. Medical Audit
   Medical Audit Advisory Committee's proposals for Making Medical Audit Effective in the District. See Appendix D. A paper for discussion and comment.
10. Total Quality Management.
    Review of chairmans attendance at last regional meeting as per attached document. See Appendix E. For information.
11. District medical committee
    Proposals for membership and election of committee. See Appendix F. For discussion and decision.
12. Management of head injury protocol
    A reminder of protocol attached. See Appendix G. For information.
13. Divisional matters
    Listed.
14. Date of next meeting
    For information.

Any Other Business should not be an item for a monthly meeting of this size. It might be acceptable for, say, Regional meetings, which perhaps only meet every six months or so, as it makes it very difficult for the Chair to control the timing of such open-ended meetings. One hospital had a meeting with only two new items on the agenda, but eleven items were brought up under Any Other Business, which doubled the length of the meeting.

The agenda should have been circulated about two weeks ahead of the meeting to allow time for the attached papers to have been read by the committee members, discussed with colleagues and pondered over. You cannot distribute documents at a meeting and expect valuable comments. A useful idea would be to group all the agenda items into routine items first, items for information only next and lastly those for discussion/decision. This enables half the agenda to be achieved quickly, giving a good sense of pace and achievement to the meetings.

There is a strong relationship between the length of time the members of a committee are given to consider items prior to the meeting and the positive decision-making process at the actual meeting. It also reduces the length of the meeting, as members are adequately prepared for what is to be discussed, and this results in a more valuable contribution to the discussion. Not least, the Chair is also fully briefed on all the topics for discussion and will have better control of the meeting, as it may be necessary to resolve any potential problems in advance.

Preparation for the meeting

As Chair, you should do your homework; there is no excuse for laxity that slows down a meeting and sends a clear message to others that the meeting is of little value. It is always very clear who has and who has not done their homework. Encourage others to do their homework as well. Never walk the group through materials page by page. The message is then clear to the others that preparation is unnecessary. Guess expectations and then exceed them. A few minutes of extra preparation will give you credibility with others.

Don't have things on the agenda if your homework has shown you cannot win. Talk to people before the meeting and do not limit your discussion to the meeting. Not only are people flattered to be approached, it will enable you to hear things you will not necessarily hear at the meeting. Sophisticated participants will be talking to each other before the meeting.

People tend to see what they want to see, picking facts to support preconceptions. Most meetings in medicine are with peers and they do not necessarily see you the way you do. Your right to control is therefore not as clear. Your authority may be resented, even challenged. The more informal your authority as leader, the tougher your job. Being blessed with the authority of a superior agent does not give the same authority as the superior.

To help you prepare, decide what you want to accomplish at the meeting. Ask yourself:

- Do I have the authority to get the job done?
- Does the group have a mandate to complete the task?
- Who is attending the meeting?
- Is there anyone you should speak with before the meeting?
- Who will be support and who will oppose the items on the agenda?
- What are the priority issues to be discussed and/or decided? And in what order?

And if you are organising a meeting for a specific purpose, which is not a routine meeting, you will also need to consider such topics as:

- Where should the meeting take place?
- When is the appropriate time?
- How or what will be the process of the meeting.
- How will decisions be reached (by consensus, implied understanding, custom and etiquette, etc.).

Distinguishing success from failure

If, after the meeting, someone asks how did the meeting go and your best reply is 'we had a good discussion', you have a problem. Every meeting is a winner or a loser. But you do need some yardstick by which to measure success or failure:

- A meeting to inform has to present and receive a predetermined amount of information.
- A brainstorming meeting is not successful unless new ideas have been generated.
- A meeting to delegate has not been successful unless each participant has a clear idea of their respective assignment.

Therefore, for each meeting, you should set a measurable objective prior to the meeting by which you can measure success or failure. The standards should, of course, be attainable.

Share the objectives and the responsibility for achieving them, put responsibility on your meeting partners for the success of the meeting. Be aware in advance what you want out of the meeting and share these objectives before the meeting with the group.

Standards for measuring meetings

Recall a recent meeting that you chaired:

- Was there a goal?
- Did you have a plan to achieve your goal?
- How succesful was the meeting?
- Was it really necessary?
- Would you do it differently another time?

Consider a meeting you have yet to attend:

- Are the goals established?
- What is your measure of success?
- After the meeting, compare your answers with what actually happened.

Chair manners

Maintain courtesy and, if you really want valuable contributions, treat all your participants with respect, some may be smarter than others, some will be more creative, others better workers and yet others more influential, but they all may have some contribution to make. Do value the introverts who do not contribute often, they may have a useful offering.

If you have a maverick in the team, or find someone irritating, try considering whether and how they are like you, or what it is about them that you really admire and would like to be like, but refuse to recognise.

The more rigorous and demanding the requirement for membership in a meeting, the more the participants will enjoy and contribute to the meeting. Demand serious preparation, attention to detail and personal effort. However, there is a difference between setting high expectations for success and effort and setting unrealistic goals. Participants need to feel like winners by achievement. Speak in terms of 'we' not 'I', especially when success is achieved. Make the task appear important; value participation; communicate confidence; even when the matter is critical, put your positive comments first; always try to look for the positive in any statement and, if you must disagree, do so with the statement, not with who says it. Congratulate the group as you move on. Last statements are often remembered, summarise actions and create expectations for future meetings. Bring a positive attitude and don't bring any of your other problems to the meeting.

Running the meeting

Begin the meeting by getting announcements out of the way first. Move to simple issues to gain momentum. The energy of the meeting diminishes as the meeting progresses. Consider more demanding issues early on if you are looking to get your own way rather than accepting serious contributions, as time dwindles endeavour to push your items through. If possible the meeting should begin and end on positive notes. The positive success at the beginning can set up a feeling of success for the remainder of the meeting. That momentum can be helpful later, so try to finish with an item that makes people feel they have accomplished something. If that is not possible, as Chair try to summarise what was positive that came out of the meeting. Set a goal. Focus everyone on that goal beforehand so that after the meeting you will be able to say clearly what has been accomplished.

Keep the meeting moving. When the issue is clearly ripe for conclusion, offer your summary, give the meeting momentum and continuing success. Clarify and summarise points of agreement and opposing points of view. People often hear things differently, your primary task is to make sure each person hears it in the same way. Take it step by step. When a particular task is completed, state the conclusion and make sure everyone is together before moving on. Make it clear what has been completed and what left open.

At the end of the meeting, summarise what has been accomplished and relate conclusions to original intentions in a positive fashion. Make the members feel their effort and time was worth it. End on time.

Dealing with problems

If someone comes late, do not draw draw attention to this by restating what has gone before or remarking on lateness; this only adds further distraction. If you know in advance that someone is coming late, note it for the group so that it does not appear as a criticism of the meeting.

Most discussions involve many tasks to reach a conclusion. Different tasks require different processes. You cannot have an effective discussion by approaching all the tasks, the problems and the people all at the same time. Separate process from structure. If part of the group is discussing structure and another part process, you will need to get the group back on the same track. Distinguishing process and structure and separating and ordering tasks are among the most critical techniques for ensuring success.

Prior to addressing the structure the group will need to know the process for making a decision. Who has decided or will decide the criteria by which the decision will be made. Will all solutions be presented before discussion. Will one or three be recommended. Will decision be by consensus, by majority or by the Chair?

Separate problems from people. What is the basic problem and the goal to be achieved? Separate the problems when there is more than one. Divide problems into subproblems that can be addressed, whenever possible, as separate issues.

Gather the facts. Remember, facts seldom speak for themselves. When dealing with people the perception of facts is more important than the facts themselves. Facts are just arguments heard in different ways, and usually dependent on who presented them.

Always ask for multiple solutions if possible, and pick the best solution. Decide who should implement the solution and make sure it gets done. There are three common problems:

(1) Defining the problem.
(2) Setting the environment for good consultation.
(3) Confusing consultation with making a decision.

This should at least be familiar ground for doctors due to the similarities with medicine itself. What seems to be the problem, i.e. if something seems to be wrong let's get it on the table for discussion. How long has it been going on, i.e. gathering the facts. Let me examine you, i.e. a detailed examination of facts, or distinguishing problems. You seem to have appendicitis, i.e. a diagnosis is offered, but you need to look for more than one (i.e. a differential diagnosis). We will operate, i.e. a solution to solve the problem, but the group should offer more than one before deciding. Finally be very careful not to let members go over old ground and begin the discussion again.

When not to hold a meeting

There may be times when you are not sure whether to hold a meeting. Do not do so if:

- You cannot define what the meeting is to accomplish.
- The meeting will not achieve its purpose by virtue of its composition or authority. In other words the meeting will be with the wrong people.
- The desired purpose could be achieved by some other means.
- You are not properly prepared or, for some reason, you cannot be properly prepared in time.

Index

Accountability
  of clinical managers 42
Agendas
  circulating 137
  examples of 138, 139
Anaesthetics 44
Audit 78–95
  change from 89
  clinical 86
  computerised 88
  confidentiality of 84
  criterion-based 87
  definition 78, 79, 85
  educational 71
  essentials 82
  global 86
  Government 91
  historical background 79
  independent 91
  involving others in 83
  key questions 82
  management and 89
  medical 85
  multispeciality 86
  national 90
  nursing 71, 73
  organisation of 82
  outcomes 81
  principles of 80
  process 81
  quality of service 87
  regional 86, 89
  scope of 81
  self- 87
  structure of 81
  surgical 86
Audit Commission 66, 91
Audit of audit 87
Audit cycle 80
Authority
  decentralisation of 26
  definitions 43
  managers 42
Autonomy
  erosion of 4

Bed occupancy
  numbers of nurses and 72
British Medical Association
  clinical directorate model 25
Budgeting
  clinical and managment 13
  clinical manager 52
  doctor's responsibility for 12
  service departments 44
Bureaucracy 31
Business manager 27, 33, 48
  finance and 51

Care 107
  holistic 69, 70
  quality of *See Quality of Care*
  teamwork in 93
  total patient 69, 70
Care plans 69
Cash limitations 12
CEPOD 79, 82, 87, 91
Central control
  relinquishing 7
Chair *See Clinical Director*
Chair of Division 23
  See also Clinical Director
Chairing meetings 132–44
  See also under Meetings
  adaptability of 134
  agenda 137, 138, 139
  anticipation and 133
  characteristics 133
  communication and 133
  control of 135
  dealing with problems 143
  examples of agenda 138, 139
  general observations 133
  managing people 133
  manners 142
  neutrality in 134
  planning 136
  preparation for 140
  presenting facts 144
  procedure 143
  reasons for holding 144

146 Index

Chairing meetings, *continued*
  role of agenda 137
  size of 156
  success or failure 141
  time factors 135
Change 96–113
  actions and awareness for 111
  agreement on plans 105
  aims and objectives and 122
  barriers to 107, 209
  bringing about 100
  committment planning 110
  consultation in 103
  decision-making and 105
  decree approach 100
  design of 103
  diagnosis of problems 102
  evaluation of 111
  external 98
  first order 99
  force field analysis 109
  group decision approach 100
  group problem-solving approach 100
  guidelines for 101
  handling 107
  implementing 80, 106
  information and 103
  in hospitals 97
  internal 98
  management of 100–110
  modification of plans 106
  need for 102
  negative forces 109
  outcome of 106
  pathways to 107, 109
  positive forces towards 109
  power and 108
  publicity 105
  replacement approach 100
  resistance to 111
  second order 99
  selection of methods 104
  sensitivity training approach 100
  solutions and alternatives 103
  steps in making 102
  structural approach to 100
  support and 106
  technical 99
  transformational 99
  transitional 99
  types of 50, 98
Change management 50
Chief *See Clinical Director*
Chief Executive Officer 23
  See also Unit General Manager
Clinical audit 42, 86
  definition of 85
Clinical budgeting 14
Clinical care 107
Clinical Director 15, 22, 26, 27, 40–58
  accountability 42
  allocation of work 36
  appointment of 32
    as coordinator 49
    as monitor 49
    as negotiator 51
  authority of 42
  budget 52
  characteristics of success 54
  clinical audit and 49
  communications 53
  competence of 46, 55
  contract 33
  dealing with conflict 50
  decision-making and 49
  demands, constraints and choices 46, 56–7
  dilemmas 57
  enterprise and initiative by 54
  financial management by 51
  informational role 48
  job description 33
  leadership 46, 54
  management of change 50
  management skills of 46
  mandate for 41
  organising communication 36
  performance appraisal 34
  personal influence 43
  quality of care and 67
  re-entry problems 34
  relationship with Business and Nurse managers 35
  relationship with Director of Nursing 64
  relationships with other managers 34, 36, 64
  relationship with UGM 33, 35, 57
  renumeration 33
  requirements for 41
  responsibility 41, 47
  responsibility for audit 82
  role of 27
  staff management 36
  succession 32
  tenure of office 32
  time employed 33
  tribalism in 45
Clinical directorates 20–31
  appointment of director 32
  associate 21
  BMA model 25
  business manager 35 *See also Business manager*
  classification of 27
  communication 36, 53
  director 32 *See also Clinical Director*
  Director of Nursing and 65
  divisions in 21
  doctor's responsibility 30
  facilities in 22

Index

Guys Hospital model 24
in-line or straight model 25
Johns Hopkins model 24
management style of 47
models of 22
nurse manager 36–7 *See also Nurse Manager*
opportunities 30
organisational structure 23
practical aspects of 26
setting up 25
sharing or transverse model 28
splitting or V model 28
team approach 56
Clinical chairs 15 *See also Clinical Directors*
Clinical freedom 13, 17, 26
management of resources and 2
Clinical management
models of 19–39
Clinical outcome 88
Clinical research 38
Clinical Standards Advisory Committee 80
Cogwheel Report and system 9, 26, 27
Colleges of health
relationship with nurses 67
Committee meetings 122 *See also Meetings*
Committment planning for change 110
Communications 53
factors involved 53
in meetings 133
Computers in nursing 75
Confidential Inquiry into Maternal Deaths 79
Conflict 50
Consultants
effect on resources 2
financial programmes and 2
in teams 93
Consultation
change and 103
Contracts management 52
Coordinator
clinical managers as 49
Corporate management 46, 53
Credibility
loss of 6
Criterion-based audit 87
Cultural change 29
Cultural differences
between doctors and managers 4

Data collection 93
Day surgery 102
Decision-making 3, 104–5
at meetings 118, 135, 143
by clinical managers 49
Director of Nursing in 63, 64
district level 5
doctors in 14
meetings for 117

organisational (neo-classical) model 104
political model 104
process (managerial) model 104
rational (classical) model 104
Delegated power 100
Delegation 118
Devolved power 7
Director of Faculty *See Clinical Director*
Director of Nursing 60
advisor and representative 63
as manager 63
asprofessional leader 62
care standards and 76
clinical directorates and 65
clinical practice and 68
computers and 75
conflict of interests and 64
developments and 73
dilemmas 76
executive 63
GPs and 67
liason with external agencies 67
Project 2000 and 74
quality of care and 67
relationships 67
resource allocation 76
responsibilities 67
role of 62, 76
Discussions
at meetings 117
District Management Teams 9
District Medical Committee 10
Divisions 9
Doctors
accepting managerial responsibilities 8
appointment to managerial positions 15
erosion of autonomy 4
erosion of power of 3
erosion of values 3
involvement in management 16
manager's helping 5
manager's view of 4
responsibility for budgets 12
role of 30
role in cost containment 11
role in management 1–18
Doctor-manager relationship 1, 8
cultural differences 4
differing values 6
misunderstandings 3
suspicions 5
terminologies 3
understanding points of view 6
Doctor-patient relationship 17

Educational audits 71
Elderly 74
Ethical values
erosion of 3
Executive Director of Nursing 63

148 Index

Family Health Service Authorities 67
Financial audit 86
Financial limitations 12
Financial management
   by Clinical Director 51
Financial programmes
   consultants and 2
Flexibility of staff 50
Future
   visions of 4, 50

General managers
   appointment of 10
   control over Clinical Directors 45
   relationship with Director of Nursing 64
General practitioners 38, 67
Global audit 86
Goals and objectives 4, 40, 48
   change and 112
Griffiths Report 1, 10, 11, 13, 15
Group dynamics 121, 123
Group mind 123
Guys Hospital
   clinical directorates in 24

Health Advisory Service 87
Health authorities 67
Health care assistants 75
Health service indicators 88
Holistic care 69, 70
Hospitals,
   doctors in management 15
   organisational change in 97
   responsibilities 30

Independent audit 91
Information
   at meetings 117
   change and 103
   doctors and 6
Information management 48
Internal marketing 45
Interpersonal roles 47
Interpersonal skills 4

Johns Hopkins Hospital 8
   clinical directorates at 24
Joint Working party on the Organisation of Medical Work in Hospitals 9

Leadership 5
   definition of 48
   delegation of 48
   of clinical managers 46, 54
   of Director of Nursing 62
Local audit Committee 84
Local Government and Housing Bill 91
Loyalty 41

Management
   audit and 89
   relation with doctors
      See Doctor-manager relationship
Management Boards 19
Management Budgeting 13, 14
Management process
   definition of 40
Managers
   as purchasers 90
   collaboration with medical staff 93
   demands, constraints and choices 56-7
   helping doctors 5
   leadership roles 46, 47
   loss of credibility 6
   pressure on 2
   relinquishing central control 7
   role of 30
   ten roles 46–7
   view of doctors 4
Matrons 59, 66
Medical Advisory Committee to Purchasers 26
Medical audit 9, 14, 85–6, 88
   definition of 85
   development of 89
   quality assurance and 88
Medical Audit Advisory Committee 82
Medical Director 37
Medical education 38
Medical Executive Committee 23
Medical Officers of Health 10
Medical staff
   collaboration with managers 93
   committees 26
Medical Superintendents 8, 10, 11
Meetings
   agenda 137, 138, 139
   brainstorming 126, 127
   chair's manners 142
   characteristics of 116
   communication in 133
   confrontation in 130
   decision-making in 117, 118, 135, 143
   defending a weak case at 130
   delegation of work or authority 118
   discussions at 117
   disguised agendas 126
   effectiveness of 120
   essentials 115
   examples of agendas 138, 139
   general techniques 129
   good and bad members 125
   ideas at 118
   information at 117
   interruption methods 129
   issues at 118
   management at 114–32
   managing chair 132–44 *See also* Chairing meetings

numbers present   121
of small groups   126
opposition in   130
people management at   133
personnel of   121
place of   127
planning   136
preparation for   122, 140
problems of   119, 128, 143
productive   119
reasons for   116
reasons for attendance   120
reasons for not holding   144
reluctant individuals at   131
role of agenda   137
scale of achievement   117
seating arrangements   127, 134
sharing responsibility   118
size of   136
social and proicess leaders   126
socialising   117, 118
standards of measurement   142
success or failure in   141
supportive   117
tensions in   124
theatre of   124
time factors in   135
time wasting at   116
types of   115
unproductive   115
voting   135
Misunderstandings   3
Models
  clinical directorates   22
  clinical management   19–39

National audit   90
National Audit office   90
National Confidential Enquiry into Perioperative deaths (CEPOD)   79, 82, 87, 91
National External Quality Assessment Scheme   79
National Health Service and Community Care Act   37
Nightingale, Florence   60
Nurses
  accountability   60
  clinical practice   68
  Code of Professional Conduct   60
  competence   61
  continuing responsibility   61
  education of   61, 74
  patient dependency and   73
  professional development   70
  work load   73
Nurse Manager   27, 33, 35, 36, 48, 60
  business managers and   66
  finance and   51
  general managers and   66
  quality of care and   67
  relationship with Director of Nursing   66
  responsibility   61, 67
  role of   62, 66
Nurse patient ratio   71, 72
Nurse practitioners   68
Nursing
  as profession   60
  care plans and models   69
  clinical grading   61
  computers and   75
  development of   60, 73
  education   70
  manpower   71
  philosophy of   68
  primary   69
  Project 2000   74
  research   70
Nursing audit   71, 73
Nursing Director *See Director of Nursing*
Nursing management   59–77
Nursing Officer   23

Openness   31
Organisations
  equilibrium of   97
Outpatients   44, 87

Patients
  collaboration with   93
Patient Care
  see under Care and Quality of Care
Patient dependency   73
Patient's Charter   67
Patients First   10
Power
  change and   100
  delegated   100
  erosion of   3
Project 2000   74

Quality of care   71, 85
  audit and   84, 87
  responsibility for   67
  Standing Medical Advisory Committee on   85

Radiology   45
Regional audit   86, 89
Report of a Confidential Enquiry into Perioperative Deaths (CEPOD)   79, 82, 87, 91
Resource allocation   76
Resource management   12, 14–5, 75
  clinical freedom and   2
  involvement of doctors in   13
Resources   12
  consultants and   2
Responsibility
  of clinical manager   47

Responsibility, *continued*
  definition 41
  of Director of Nursing 67
  division of 2
  of meetings 118
  of nurse managers 61, 67
  of ward sisters 61
Royal College of General Practitioners 79
Royal College of Physicians 79
Royal College of Surgeons of England 78, 86, 91
Royal Commission 10

Self-audit 87
Senior Nursing officer 23
Service departments
  budgets of 42
Service manager 52
Skill in management 7
Staff
  communication with 53
  motivation and goodwill 57
Standing Medical Advisory Commitee of Quality of Medical Care 85
Surgical audit 86

Technical change 99

Technical competence 55
Terminology
  differing 3
Theatres 44
Total patient care 69
Total Quality Management 92
Tribalism
  in clinical managers 45
Trusts 676
  Director of Nursing and 63
  relations with GPs 38
Trust boards 19

Unit General Manager 23, 24, 37
  Clinical Directors and 33, 35, 57
  Nurse Managers and 66
  Unit Management Teams 10

Values
  differences in 6
  erosion of 3

Waiting lists 44
Ward sisters 68, 70
  responsibility of 61
Working for Patients 11, 37